GENETICS IN THE WORKPLACE
Implications for Occupational Safety and Health

Department of Health and Human Services
Centers for Disease Control and Prevention
National Institute for Occupational Safety and Health
Genetics Working Group

ORDERING INFORMATION

To receive NIOSH documents or more information about occupational safety and health topics, please contact NIOSH:

>Telephone: 1-800-CDC-INFO (1–800–232–4636)
>TTY:1–888–232–6348
>E-mail: cdcinfo@cdc.gov

>or visit the NIOSH Web site at www.cdc.gov/niosh

For a monthly update on news at NIOSH, subscribe to *NIOSH eNews* by visiting www/cdc.gov/niosh/eNews.

This document is in the public domain and may be freely copied or reprinted

Disclaimers

Mention of any company or product does not constitute endorsement by NIOSH.

Citations to web sites external to NIOSH do not constitute NIOSH endorsement of the sponsoring organizations or their programs or products. Furthermore, NIOSH is not responsible for the content of these web sites.

DHHS (NIOSH) Publication Number 2010-101

November 2009

FOREWORD

The purpose of this document is to consolidate the diverse literature and opinions on genetics in the workplace, to flag important issues, and to provide some considerations for current and future research and practice. Recent advances in understanding the human genome have created opportunities for disease prevention and treatment. Even though the focus of attention on applications of genetic discoveries has been largely outside of the workplace, genetic information and genetic testing are impacting today's workplace.

The issues related to genetic information and genetic testing in the workplace have the potential to affect every worker in the United States. This NIOSH document provides a discussion on the benefits, limitations, and risks of genetic information and genetic tests. Anecdotal evidence already exists of employers inappropriately using genetics tests. Although genetic technology is becoming widely available, a serious knowledge gap on the part of consumers of this technology is a concern. Basic information on genetics, genetic research, genetic testing, genetic information, informed consent, privacy, confidentiality, technological advances based on genetics, notification, data management, and discrimination need to be discussed. The passage of the Genetic Information Nondiscrimination Act of 2008 has abated some concerns about the misuse of genetic information. This NIOSH document provides information on these issues to help the reader be made more aware of the multitude of scientific, legal, and ethical issues with regard to the use of genetics in occupational safety and health research and practice.

This document has been written to appeal to both targeted and broad audiences. Occupational safety and health professionals and practitioners interested in the use of genetic information in the workplace will be most informed by the chapters on the role of genetic information in the workplace, health records, genetic monitoring, genetic screening, and the ethical, social, and legal implications of this information. Academics and researchers will be especially interested in the chapter on incorporating genetics into occupational health research. Employers, workers, and other lay readers will likely find the chapters on health records and ethical, social, and legal implications of genetic information in the workplace provide the most information. Regardless of specific reader interest levels, the goal of this document is to draw attention to the many gaps in knowledge about the use of genetic information and to stimulate dialogue on its use in the workplace.

John Howard, M.D.
Director
National Institute for Occupational Safety and Health
Centers for Disease Control and Prevention

GENETICS WORKING GROUP

D. Gayle DeBord, Ph.D., Chair

Paul Schulte, Ph. D.

Mary Ann Butler, Ph.D.

Erin McCanlies, Ph.D.

Susan Reutman, Ph.D.

Avima Ruder, Ph.D.

Anita Schill, Ph.D.

Mary Schubauer-Berigan, Ph.D.

Christine Schuler, Ph.D.

Ainsley Weston, Ph.D.

ACKNOWLEDGMENTS

Editorial Review by Robert J. Tuchman

Desktop Publishing and Camera Copy Production by Brenda J. Jones

Web Development by Julie Zimmer

Table of Contents

GENETICS IN THE WORKPLACE i

ORDERING INFORMATION ii

FOREWORD iii

GENETICS WORKING GROUP iv

ACKNOWLEDGMENTS iv

EXECUTIVE SUMMARY x

GLOSSARY xv

INTRODUCTION 1

THE ROLE OF GENETIC INFORMATION IN OCCUPATIONAL DISEASE 5

2.1 Gene-Environment Interactions in Occupational Safety and Health Research 5
2.2 The Effect of Occupational and Environmental Exposures on Genetic Material 13
2.3 Effect of Genes on Biomarker Measurements and Use in Risk Assessments 14
2.4 Genomic Priorities and Occupational Safety and Health 14

INCORPORATING GENETICS INTO OCCUPATIONAL HEALTH RESEARCH 17

3.1 Validity and Utility Issues of Genetic Assays 19
3.2 Challenges of Genomics and Related Research Areas 24
3.3 The Relationship Between Genetic and Environmental Risk Factors 27
3.4 Analytical Epidemiological Research 30
3.5 Use of Banked or Stored Specimens 35
3.6 Cell Lines and Transgenic Animals 35
3.7 Considerations in the Incorporation of Genetics into Occupational Health Research 36

HEALTH RECORDS: A SOURCE OF GENETIC INFORMATION ... 39

- 4.1 Health Inquiries and Examinations ... 40
- 4.2 Confidentiality, Privacy, and Security ... 40
- 4.3 Genetic Exceptionalism ... 42
- 4.4 Genetic Discrimination ... 43
- 4.5 The Historical Use of Genetic Information ... 43
- 4.6 The Use of Genetic Information in Research ... 44
- 4.7 Genetic Information in the Assessment of Causation ... 44

GENETIC MONITORING: OCCUPTIONAL RESEARCH AND PRACTICE ... 47

- 5.1 Regulation ... 50
- 5.2 Considerations for Genetic Monitoring ... 51

THE THEORETICAL USE OF GENETIC SCREENING AND OCCUPATIONAL HEALTH PRACTICE ... 53

- 6.1 History ... 54
- 6.2 Past and Current Use of Genetic Screening ... 55
- 6.3 Technical and Public Health Issues in Worker Screening ... 57

THE ETHICAL, SOCIAL, AND LEGAL IMPLICATIONS OF GENETICS IN THE WORKPLACE ... 61

- 7.1 Framework for Considering Genetic Information ... 62
- 7.2 Inherited Genetic Factors: Research ... 62
- 7.3 Inherited Genetic Factors: Practice ... 66
- 7.4 Inherited Genetic Factors: Litigation and Regulation ... 72
- 7.5 Acquired Genetic Effects: Research ... 74
- 7.6 Acquired Genetic Effects: Practice ... 75
- 7.7 Acquired Genetic Effects: Regulation and Litigation ... 75
- 7.8 The Adequacy of Safeguards to Protect Workers Against Misuse of Genetic Information ... 77

REFERENCES ... 81

WEB SITES FOR FURTHER INFORMATION ... 115

INDEX ... 117

ABBREVIATIONS AND ACRONYMS

AAOHN	American Association of Occupational Health Nurses, Inc.
ACCE	analytic validity, clinical validity, clinical utility, and ethical, legal, and social implications
ACE	angiotensin I converting enzyme (peptidyl-dipeptidase A) 1
ACMG	American College of Medical Genetics
ACOEM	American College of Occupational and Environmental Medicine
ADA	Americans With Disabilities Act of 1990
ADH	alcohol dehydrogenase
ADRB3	adrenergic, beta-3-, receptor
ALAD	aminolevulinate, delta-, dehydratase
AMA	American Management Association
APOε	apolipoprotein E
AR	androgen receptor
ASHG	American Society of Human Genetics
BMI	body mass index
BRCA2	breast cancer gene 2, early onset
CBD	chronic beryllium disease
CCND1	cyclin D1
CDKN2A	cyclin-dependent kinase inhibitor 2A
CDC	U.S. Centers for Disease Control and Prevention
CFR	Code of Federal Regulations
CLIA	Clinical Laboratory Improvement Amendment of 1988
CNV	Copy number variant
CTSD	cathepsin D
CYP1A1	cytochrome P450, family 1, subfamily A, polypeptide 1
CYP1B1	cytochrome P450, family 1, subfamily B, polypeptide 1
CYP17	cytochrome P450, family 17
CYP2D6	cytochrome P450, family 2, subfamily D, polypeptide 6
CYP2E1	cytochrome P450, family 2, subfamily E, polypeptide 1
CYP3A4	cytochrome P450, family 3, subfamily A, polypeptide 4
DHHS	U.S. Department of Health and Human Services
DNA	deoxyribonucleic acid
DOE	U.S. Department of Energy
DPB1	major histocompatibility complex, class II, DP beta 1
DQ	major histocompatibility complex, class II, DQ
DQA1	major histocompatibility complex, class II, alpha 1
DQB1	major histocompatibility complex, class II, beta 1
DR3	major histocompatibility complex, class II, DR3
DRB1	major histocompatibility complex, class II, DR beta 1
EEOC	Equal Employment Opportunity Commission

EGP	Environmental Genome Project
ERα	Estrogen receptor α (*ESR1*)
ERβ	Estrogen receptor ß (*ESR2*)
ERCC2	excision repair cross-complementing rodent repair deficiency, complementation group 2 (xeroderma pigmentosum D)
FDA	U.S. Food and Drug Administration
FISH	fluorescence in situ hybridization
G6PD	glucose-6-phosphate dehydrogenase
GEI	Genes, Environment and Health Initiative
GINA	Genetic Information Nondiscrimination Act of 2008
GJB2	gap junction protein, beta 2, 26kDa (connexin 26)
GPA	glycophorin A
GST	glutathione S-transferase
GSTM1	glutathione S-transferase mu1
GSTP1	glutathione S-transferase pi
GSTT1	glutathione S-transferase theta 1
GWAS	genome-wide association studies
HIPAA	Health Insurance Portability and Accountability Act
HIV	human immunodeficiency virus
HLA	human leukocyte antigen
hMLH1	mutL homolog 1, colon cancer, nonpolyposis type 2 (E. coli)
hMSH2	mutS homolog 2, colon cancer, nonpolyposis type 1 (E. coli)
HPRT	hypoxanthine phosphoribosyltransferase
HuGE	Human Genome Epidemiology
HuGENet	Human Genome Epidemiology Network
IARC	International Agency for Research on Cancer
IL-3	interleukin 3
IL-4 and 4R	interleukin 4, interleukin 4 receptor
IRB	institutional review board
Ki-ras	Kirsten rat sarcoma viral oncogene homolog
MCP-1	monocyte chemoattractant protein-1
MIAME	minimum-information-about-a-microarray experiment
MnSOD2	manganese superoxide dismutase 2
MPO	myeloperoxidase
mRNA	messenger ribonucleic acid
MTHFR	5,10-methylenetetrahydrofolate reductase (NADPH)
MTR	5-methyltetrahydrofolate-homocysteine methyltransferase
NAS	National Academy of Sciences
NAT1	N-acetyltransferase 1 (arylamine N-acetyltransferase)
NAT2	N-acetyltransferase 2 (arylamine N-acetyltransferase)
NBAC	National Bioethics Advisory Commission
NCI	U.S. National Cancer Institute

NF1	neurofibromin 1 (neurofibromatosis, von Recklinghausen disease, Watson disease)
NF2	neurofibromin 2
NIEHS	U.S. National Institute of Environmental Health Sciences
NIH	U.S. National Institutes of Health
NIOSH	U.S. National Institute for Occupational Safety and Health
NQO1	NAD(P)H dehydrogenase, quinone 1
NRC	National Research Council
OGG1	8-oxoguanine DNA glycosylase
OSH Act	Occupational Safety and Health Act of 1970
OSHA	U.S. Occupational Safety and Health Administration
OTA	Office of Technology Assessment
PAH	polycyclic aromatic hydrocarbons
PCR	polymerase chain reaction
PGR	progesterone receptor
PIR6.2	potassium inward-rectifier 6.2
PON1	paraoxonase 1
PPAR-γ	peroxisome proliferative activated receptor, gamma
PPV	positive predictive value
PTEN	phosphatase and tensin homolog
RFLP	restriction fragment length polymorphism
RNA	ribonucleic acid
ROC	relative operating characteristic
SACGT	Secretary's Advisory Committee on Genetic Testing
SD	standard deviation
SNP	single nucleotide polymorphism
SP-B	surfactant protein B
TDI	toluene diisocyanate
TGFB1	transforming growth factor beta 1 - induced
TGFBR2	transforming growth factor beta receptor II
TNF-α	tumor necrosis factor (TNF superfamily, member 2)
TNF-α-308	tumor necrosis factor (TNF superfamily, member 2 with a base change at position 308
TP53	tumor protein p53
tRNA	transfer ribonucleic acid
UGT2B7	UDP-glucuronosyltransferase 2 family, polypeptide B7
Val	valine
XRCC1	X-ray repair complementing defective repair in Chinese hamster cells 1
XRCC3	X-ray repair complementing defective repair in Chinese hamster cells 3
XRCC5	X-ray repair complementing defective repair in Chinese hamster cells 5

EXECUTIVE SUMMARY

> - Exposure to a workplace hazard is necessary for an occupational disease to occur, regardless of the genetic makeup of the person.
> - The use of genetic information in occupational safety and health research and practice would have several real or perceived consequences.
> - In occupational safety and health practice, genetic tests—whether for monitoring or screening—must be validated to provide reliable exposure or risk assessments.
> - At this time, no genetic test related to an occupational disease has been validated or accepted for use, except the use of genetic biomarkers to measure the dose of a genotoxic exposure.

Major technological advances in the last few years have increased our knowledge of the role that genetics has in occupational diseases and our understanding of genetic components and the interaction between genetics and environmental factors. The use of genetic information, along with all of the other factors that contribute to occupational morbidity and mortality, will play an increasing role in preventing occupational disease. However, the use of genetic information in occupational safety and health research and practice presents both promise and concerns [McCanlies et al. 2003; Kelada et al. 2003; Henry et al. 2002; Christiani et al. 2001; Schulte et al. 1999]. Use of genetic information raises medical, ethical, legal and social issues [Clayton 2003; Ward et al., 2002; McCunney 2002; Christiani et al 2001; Rothstein 2000a; Schulte et al. 1999; Lemmens 1997; Barrett et al. 1997, Van Damme et al. 1995; Gochfeld 1998; Omenn 1982].

The purpose of this report is to bring together the diverse literature and opinions on genetics in the workplace, to highlight important issues, and to provide some considerations for current and future practice. Occupational safety and health professionals and practitioners may have particular interest in this report as the understanding of gene-environment interactions at the mechanistic and population levels may result in improved prevention and control strategies. This report is divided into topic areas for ease of reading. Specifically, the role of genetic information in occupational disease is discussed in Chapter 2, followed in Chapter 3 by a presentation of how genetics is incorporated into occupational health research. Health records as a source of genetic information are discussed in Chapter 4. The report continues in Chapter 5 with a focus on genetic monitoring, followed in Chapter 6 by a theoretical discussion of genetic screening. The final chapter presents an overview of the most important aspects of this report, which are the ethical, social, and legal implications of genetics in the workplace. In addition, ethical issues specific to health records and genetic testing are discussed in Chapters 4 and 6, respectively. To assist our audience in finding additional sources of information or more in-depth discussion of the issues surrounding genetic information, a list of web sites is provided at the conclusion of this document.

Role of Genetic Information in Occupational Disease. The role of genetic information in occupational disease is being explored. The framework for considering genetics in the exposure to disease paradigm arose from a National Academy of Sciences review on biomarkers [NRC 1987]. Biomarkers are measurements using biological tissues that give information about exposure, effect of exposure, or susceptibility. Evaluation of genetic damage can provide information about exposure or effect of exposure. However, the presence of a specific genetic biomarker will not itself result in an occupational disease; exposure to a workplace hazard is necessary. The presence of a disease risk biomarker in the absence of exposure may be innocuous.

The study of biomarkers of genetic susceptibility in the context of workplace exposures can provide information about gene-environment interactions. One major emphasis of genetic research in occupational disease has been in the area of response variability.

Extensive variability in the human response to workplace exposures has been observed. Genes can have multiple variations known as polymorphisms, which may contribute to some of this variability [Grassman et al. 1998]. Research has been conducted over the last approximately 30 years to identify the role of genetic polymorphisms in a wide range of occupational and environmental diseases, particularly those involving occupational carcinogens [Hornig 1988; Berg 1979]. The risk of biological effects or diseases attributable to an occupational exposure can be decreased, unchanged, or increased among individuals with certain genetic polymorphisms.

Incorporating Genetics into Occupational Health Research. The main influence on genetic research with respect to occupational health is the large number of technological advances in molecular biology. Because of these new techniques, it is now feasible to evaluate the relationship of disease with individual genes and their variants or even with the whole genome. These technologies promise to set the stage for new discoveries in understanding mechanisms and the preclinical changes that might serve as early warnings of disease or increased risk [NRC 2007]. They also present difficult challenges in terms of handling large data sets, understanding the normal range, standardizing technologies for comparison and interpretation, and communicating results [King and Sinha 2001; Wittes and Friedman 1999].

As our understanding of the role of specific genes and their variants increases, genetic tests are being developed to look at specific genotypes. One critical issue in genetics is the validity of such genetic tests. Much contemporary genetic research involves the collection of biological specimens (usually DNA in white blood cells) that are then tested either for changes (damage) to genetic material or for genetic polymorphisms. These genetic tests, while useful in occupational health research, are not ready for clinical use; in other words, they are not validated for clinical interpretation. Validation is a process by which a test's performance is measured both in the laboratory and in populations, resulting in the evaluation of the clinical utility or the risks and benefits of the test. Until clinically validated, the information from such tests may be meaningless with regard to an individual's health or risk. In contrast, genetic tests may be validated for assessing exposure or effect modification in research even if they have no clinical utility.

Health Records: A Source of Genetic Information. Genetic tests are not the only source of genetic information in the workplace. Genetic information is kept in workers' personnel and workplace health records [Rothenberg et al. 1997]. This information is in the family history of diseases with known strong genetic etiologies as well as in the results of physical examinations and common laboratory tests. This type of information is reported routinely by workers or obtained by employers from workers' job applications, health questionnaires, health and life insurance applications, physicals, and workers' compensation proceedings [Anderlik and Rothstein 2001]. The line between what is and is not genetic information in health records is unclear. States have enacted legislation with widely varying definitions of what constitutes genetic information from an employee's health record. Questions concerning confidentiality, privacy and security remain as the handling of health records may be influenced by various federal and state regulations.

Genetic Monitoring and Occupational Research and Health Practice. Genetic information can be a scientific tool to understand mechanisms and pathways in laboratory research and as independent or dependent variables in population research studies of workers. In occupational safety and health practice, genetic tests may be used in a variety of ways. As in other areas of health science, genetic information may be used in the differential diagnosis of disease, allowing clinicians to consider or exclude various diagnoses. Monitoring for the effects of exposure on genetic material, such as chromosomes, genes, and constituent deoxyribonucleic acid (DNA), has been used to evaluate risks and potential health problems for more than 50 years, particularly those from ionizing radiation [Mendelsohn 1995; Langlois et al. 1987; Berg 1979]. Such monitoring is not unlike monitoring for metals in blood, solvents in breath, or dusts in lungs and presents less ethical concern than assessing heritable effects [Schulte and DeBord 2000]. Tests for genetic damage have been advocated as a way to prioritize exposed individuals for more thorough medical monitoring [Albertini 2001].

Genetic monitoring highlights the confusion that exists between individual and group risk assessment. Unlike other monitoring methods, the risks linked to cytogenetic changes are interpretable only for a group, not for an individual [Schulte 2007; Murray 1983; Lappe 1983].

Currently, no U.S. regulations exist that mandate genetic monitoring. Questions arise whether genetic monitoring indicates a potential health problem, an existing health problem, or compensable damage. More research is needed to understand the science before the individual's risk of disease can be interpreted from genetic monitoring results. However, genetic monitoring to determine exposure may be useful for the occupational health practitioner.

Theoretical Use of Genetic Screening and Occupational Health Practice. Genetic monitoring may have some application in occupational health practice, but perhaps the most controversial use of genetic information would be in making decisions about employment opportunities and health and life insurance coverage [Schill 2000; Bingham 1998; Van Damme et al. 1995; Murray 1983; Lappe 1983]. This would occur primarily as a result of genetic screening, in which a job (and insurance) applicant or a current worker might be asked to undergo genetic testing to determine if he or she has a certain genotype. However, the Genetic Information Nondiscrimination Act of 2008 (GINA) prohibits discrimination on the basis of genetic information with respect to health insurance and employment [U.S. Congress 2008]. Genetic screening which was not strictly prohibited by the Americans With Disabilities Act of 1990 (ADA) is now prohibited. Under ADA an employer may not make medical inquiries about an applicant until a conditional offer has been extended. Once the offer has been tendered, an employer could have obtained medical, including genetic, information about a job applicant. ADA did not prohibit obtaining genetic information or genetic screening, nor did it prohibit an employer from requesting genetic testing once an applicant has been hired provided the testing is job related. and can be used for the purposes of job placement after a conditional job offer is made.

Most criteria for genetic screening programs indicate that participation should be voluntary, with informed consent in place. Genetic screening for these purposes cannot be supported at this time because of the current lack of linkage of causation of a given genetic polymorphism with a given occupational disease and its implications with regard to the Occupational Safety and Health Act (OSH Act of 1970) that the workplace be safe for all workers [OSHA 1980b].

Accurate genetic screening information may eventually be useful to workers considering employment options. Obtaining this information for the worker would become appropriate only after the screening tests have been validated regarding risk. Various ethical arguments have been advanced in the discussion of genetic screening, and a broad range of implications of genetic testing has been discussed in the literature and in this document, including the oversight of genetic testing laboratories.

The Ethical, Social, and Legal Implications of Genetics in the Workplace. A concern about the use of genetic information in occupational safety and health is that the emphasis in maintaining a safe and healthful workplace could shift from controlling the environment to excluding the vulnerable worker. This would be counter to the spirit and the letter of the OSH Act of 1970 [OSHA 1980b]. Actions that attempt to depart from providing safe and healthful workplaces for all should not be supported. Nevertheless, understanding the role of genetic factors in occupational morbidity, mortality, and injury is important and could lead to further prevention and control efforts. However, occupational safety and health decision-makers, researchers, and practitioners may find that genetic factors do not contribute substantially to some occupational diseases. Environmental risk factors will probably always be more important for developing strategies for prevention and intervention in occupational disease and ultimately for the reduction of morbidity and mortality. The challenge is to identify and apply genetic information in ways that will improve occupational safety and health for workers.

The use of genetic information in occupational safety and health research requires careful attention because of the real or perceived opportunities for the misuse of genetic information. Society in general and workers in particular have concerns that discrimination and lack of opportunity will result from the inappropriate use of genetic information [MacDonald and Williams-Jones 2002; Maltby 2000]. While only sparse or anecdotal information supports this contention, a wide range of workers, legislators, scientists, and public health researchers have concern that such discrimination could occur. Thus, GINA and other regulations were passed to prevent the potential misuse and abuse of genetic information in the workplace. Examples of safeguards include rules and practices for maintaining privacy and confidentiality, prohibition of discrimination, and support of a worker's right of self-determination (autonomy) with regard to genetic information.

Many of these safeguards have been built into biomedical research in general, and occupational safety and health research in particular, through guidance given in the Nuremberg Code [1949], the Belmont Report [1979], and the Common Rule (45 CFR* 46) [DHHS 2005; CFR 2007], as well as in the National Bioethics Advisory Commission (NBAC) reports [1999], the ADA [1990], Health Insurance Portability and Accountability Act (HIPAA) [1996] and GINA in 2008 [U.S. Congress 2008]. ADA and HIPAA provided some safeguards against the potential misuse of genetic information in the workplace before GINA was signed and in 2000, Executive Order 13145 was signed that prohibits discrimination in federal employment based on genetic information [65 Fed. Reg.† 6877 (2000)].

In summary, the use of genetic information in the workplace has the potential to affect every worker in the United States. This NIOSH document provides information on the scientific, legal, and ethical issues with regard to the use of genetics in occupational safety and health research and practice.

---

*Code of Federal Regulation. CFR in references.
†Federal Register. See Fed. Reg. in references.

GLOSSARY

Allele: The abbreviation for allelomorphs, meaning different forms of a gene or specific DNA (deoxyribonucleic acid) sequence that may be found in a population [Jorde et al. 1997; Last 1988].

Analytical validity: The measure of how well a test predicts the disease genotype [SACGT 2000].

Base pair: Two complementary nucleotides linked by weak electrostatic bonds (hydrogen bonds). Electrostatic bonds between complementary single strands of DNA maintain the double helix structure of DNA [SACGT 2000].

Bioinformatics: The biotechnology revolution has created enormous quantities of data, especially in the areas of genomics, transcriptomics, and proteomics. Bioinformatics is the computer-based management, integration, and analysis of biotechnology data.

Biological variability: Variation in biological measurements. It can be subdivided into intraindividual variability, which is the difference within an individual over time, and interindividual variability, which is the difference between individuals.

Biologically effective dose: The amount of a substance/chemical that reaches the target site for toxicity/disease.

Biomarker: Measurement made in biological tissues that give information about exposure, disease or susceptibility.

Centromere: The condensed or constricted part of a chromosome, also known as the primary constriction. This is the structure responsible for chromosomal attachment to the spindle fibers during cell division that ensures that each daughter cell receives exactly half of the chromosomes.

Chemical base: An essential building block of DNA, one of four bases or nucleotides (adenine [A], cytosine [C], guanine [G], and thymine [T]) [SACGT 2000].

Chromosomal aberration: Structural alteration of the chromosomes. Aberrations include breaks, deletions, insertions, translocations (of part of one chromosome to another chromosome), missing chromosomes (e.g., Turners), and extra chromosomes (e.g., trisomy).

Chromosome: Literally, "colored body" (Greek). A nucleic acid-protein complex containing a DNA molecule that codes for various genes. Humans have 23 pairs of chromosomes and inherit one of each pair from each parent.

Clinical utility: Assessment of the risks and benefits of a test to determine its value or usefulness for disease prevention, disease treatment, or life planning.

Clinical validity: The measure of how well a test predicts the disease phenotype, including parameters such as positive predictive value and penetrance, environmental factors, and prevalence of the condition [SACGT 2000; Khoury et al. 1985].

Confounding: In epidemiology, a distortion of the exposure-disease relationship due to the effect of another variable (the confounder). A confounder must (1) be a risk factor for the disease, even among those not exposed, (2) be associated with the primary exposure of interest in the population from which the disease cases arose, and (3) not be an intermediate step in the causal pathway between the primary exposure of interest and the disease. Control of confounding, through study design or data analysis strategies, will reduce the impact of this bias [Rothman and Greenland 1998].

Copy number variant (CNV): A segment of DNA for which copy number differences have been observed when comparing two or more genomes. Normally, humans have two copies of each autosomal region. CNV can be caused by inversions, deletions, duplications, and translocations; and may be inherited or occur due to environmental exposures. CNV have been associated with a number of diseases.

Deoxyribonucleic acid (DNA): The hereditary material, a polymeric macromolecule composed of sugar, phosphate, and the four chemical bases. The ordering of the bases provides the chemical code that governs all vital processes.

Diagnostic test: A tool used to ascertain current disease status.

DNA adduct: When a chemical attaches or binds to DNA or a DNA base modified by the covalent addition of an electrophile [Poirier and Weston 2002].

DNA strand break: The usual structure of DNA is a pair of complementary strands entwined in the form of a double helix. DNA damage can result from cleavage of one or both strands, either through interaction with toxic agents (chemicals or radiation) or through interaction with enzymes that modify DNA.

Effect modification: In epidemiology, a change in the size or magnitude of the exposure-disease relationship according to the value of another variable (the effect modifier). It is not a bias to be eliminated, but a finding to be reported as part of the effect measure.

Electrophile: In chemistry, is a chemical attracted to electrons and by accepting electrons from another chemical allows it to bond or attach to that chemical.

Epigenetics: The study of heritable changes in gene function that occur without a change in the sequence of DNA.

Gene: The fundamental physical and functional unit of heredity. A gene is an ordered sequence of nucleotides located in a particular position on a particular chromosome that encodes a specific functional product (i.e., a protein or ribonucleic acid [RNA] molecule).

Gene-environment interaction: The combination of environmental factors and genetic factors to bring about a biological effect (disease phenotype). Six possible interaction models have been described: (1) neither the gene alone nor the exposure alone will result in disease, (2) a benign genotype occurs in the absence of a specific toxic exposure, (3) the risk of disease is higher in the presence of a specific genotype regardless of any type of exposure, although the risk of disease may be exacerbated in the presence of a specific exposure, (4) either the gene alone or the exposure alone may result in disease, (5) the presence of the gene decreases the effect of the exposure, and (6) the presence of the exposure decreases the effect of the gene [Khoury et al. 1993].

Gene-gene interactions: Interactions that occur when the combined effect of having certain alleles for specific genes is additive or multiplicative.

Genetic discrimination: The use of genetic information to exclude certain individuals from opportunities for employment, insurance, or reproduction.

Genetic exceptionalism: The view that genetic information can and should be differentiated from other health information and afforded special protection [Kulynych and Korn 2002; Murray 1997].

Genetic information: Narrowly defined, the results of DNA analysis; broadly defined, family health histories and the results of common laboratory tests for gene products as well as DNA test results [Kulynych and Korn 2002].

Genetic monitoring: The periodic evaluation of an exposed population, ascertaining whether an individual's genetic material has been altered over time and thus indicating exposure and/or providing an early warning of possible health effects or outcomes [Schulte and Halperin 1987].

Genetic research: Evaluation of the role of specific genes in human disease or the effect of agents on DNA material or changes in gene expression. In the context of this publication, genetic research evaluates the role of specific genes, genetic damage, or changes in gene expression in occupational disease.

Genetic screening: A panel of genetic tests performed on one individual or a single genetic test applied in a population-based program [Press and Burke 2001; Parker and Majeske 1996].

Genetic test: An assay used for the determination of the inherited genotype to identify specific alleles or mutations. Genetic tests can also be used for the determination of somatic changes that occur in the DNA, e.g., chromosomal breaks, rearrangements, or mutations.

Genome: The full complement of DNA that encodes all of the genes required for the structure and function of an organism. It has been estimated that about 20,000–25,000 human genes exist [Clamp et al. 2007].

Genome-wide association study: An approach that involves rapidly scanning markers across complete sets of DNA or genomes of many people to find genetic variations associated with a particular disease.

Genomics: The study of genomes.

Genotype: The genetic constitution of an organism, as distinguished from its physical appearance or phenotype [DOE 2001]. The distinct set of alleles that an individual carries [SACGT 2000].

Hardy-Weinberg equilibrium: The principle that, in an infinitely large population and in the absence of mutation, migration, selection, and nonrandom mating, both gene and genotype frequencies are balanced according to the formula $p^2 + q^2 + 2pq = 1.00$, where p is all of the alleles that are homozygous and q is the frequency of other alleles at a genetic locus. Many loci have greater than two polymorphic alleles.

Heterozygous: Refers to an individual who has inherited different alleles at the same genetic locus.

Hierarchy of controls: The principle that the best way to reduce worker exposure is to use a number of approaches that are designed to eliminate or minimize exposure. These include, in order of preference, elimination or substitution of the hazard, engineering controls (such as isolation or ventilation), administrative controls, training in work practices, and personal protective equipment [Ellenbecker 1996].

Homozygous: Refers to an individual with two identical copies of a gene or an individual who has inherited identical alleles of a gene.

Linkage disequilibrium: Between genes close together on a chromosome, the co-inheritance of particular alleles at each locus.

Locus (plural, loci): The position on a chromosome of a gene or other chromosome marker; also, the DNA at that position. The use of locus is sometimes restricted to mean expressed DNA regions.

Medical monitoring: The ongoing performance and analysis of routine environmental or clinical measurements aimed at detecting changes in the environment or health status of an individual or specific population that are at a known risk for a disease [Last 1988].

Medical removal: Removal of a worker from a particular job task, title, worksite, or employment by an employer due to (1) a present medical condition or perceived susceptibility to a future medical condition believed to affect workplace performance negatively or (2) an exposure believed to negatively affect the worker's current or future health or that of his/her children.

Medical screening: The presumptive identification of unrecognized disease or defect by the application of tests, examinations, or other procedures that can be applied rapidly. Screening tests distinguish apparently well persons who probably have a disease from those who probably do not and are a snapshot of the health of a person. A screening test is not intended to be diagnostic, although medical screening is performed for a clinical approach. Persons with positive or suspicious findings must be referred to their health care providers for diagnosis [Last, 1988].

Medical surveillance: Medical surveillance in the workplace is the overall process of ongoing, systematic collection, analysis, and interpretation of health data from workers. The goal is early recognition of trends and adverse health effects for prevention efforts. Included in medical surveillance is medical monitoring and medical screening.

Metabonomics: Investigations into the genetic underpinnings of metabolism.

Microarray: A set of microscopic molecular probes (nucleic acids or proteins/antibodies) that may be used to identify and quantify DNA, RNA, or proteins.

Micronuclei: Remnants of aberrant cell division that form small nuclear bodies that may contain chromosomes or chromosome fragments.

Monogenic disease: A disease determined by an allele at a single genetic locus.

Mutation: A change in the genetic material. Only changes in germ cells (ova and sperm) can be inherited. Changes in somatic cells (any cell in the body except germ cells) may lead to cancer.

Negative Predictive Value: The proportion of individuals who test negative and who will not get the condition.

Oncogene: An activated protooncogene. Oncogenes can promote or allow the uncontrolled growth of cancer cells.

Outlier: A test result quite different from the usual, typically a result that is more than two standard deviations from the mean of population test results.

Penetrance: The proportion of a population with a disease-related genotype that displays the phenotype (e.g., that develops the disease).

Phase I enzymes: Enzymes with broad substrate specificity that have the primary function of detoxification, usually by the addition of an oxygen molecule (e.g., cytochrome P450). However, these enzymes can inadvertently activate otherwise chemically inert toxincant to powerful electrophiles capable of forming adducts to DNA.

Phase II enzymes: Enzymes with narrow substrate specificity that reduce (epoxide hydrolase) or conjugate (glucuronosyl transferase) oxygenated chemical species. The primary function of phase II enzymes is detoxification; however, many have roles in the activation of procarcinogens.

Phenotype: The physical and biochemical characteristics of individuals as determined by their genotypes and by environmental factors [SACGT 2000].

Polygenic disease: A disease determined by alleles at multiple genetic loci.

Polymerase chain reaction (PCR): The polymerase chain reaction is a method for the rapid amplification of specific segments of DNA. Specificity is accomplished through the use of primer molecules that are complementary to sequences that flank the region of DNA to be amplified. The reaction mixture is then subjected to thermal cycling. Since each new cycle produces more DNA as a template, the PCR product accrues exponentially. Thus, from 1 picogram (10^{-12} grams) of original template, 25 cycles will realize more than 10 micrograms (10^{-5} grams) of PCR product.

Polymorphism: Literally, "many forms" (Greek). A locus where two or more alleles have gene frequencies greater than 0.01 in a population. For a given population, therefore, if at least 1% of genes (1 heterozygote in 50 people) harbor a DNA sequence variation, the variation is considered to be a polymorphism [Jorde et al. 1997].

Positive predictive value (PPV): The positive predictive value of a screening test is the proportion among those who test positive who truly are positive (true positives/(true positives + false positives)).

Protein adduct: An amino acid modified by covalent interaction with an electrophile.

Protein expression: The process by which transfer RNA (tRNA) translates messenger RNA (mRNA) to form proteins from amino acids.

Proteome: The full complement of proteins required for the structure and function of an organism.

Proteomics: The study of proteomes.

Protooncogene: Normal cell cycle regulatory genes that preserve cellular homeostasis. Mutations in protooncogenes can cause their activation, transforming them into oncogenes.

Reporter gene: In general, a reporter gene can give information about exposure, effect of exposure, or the expression of genes. Specific examples include (1) a gene in which mutations may be detected that signify carcinogenic or mutagenic exposure (hypoxanthine phosphoribosyltransferase [HPRT] and glycophorin A [GPA]), (2) a gene in which specific mutations suggest exposure to a specific carcinogen (codon 279 mutations of *p53* in aflatoxin B1-induced liver cancer), (3) a gene whose expression is induced in response to a specific environmental stimulus (alcohol induction of hepatic cytochrome P450, family 2, subfamily E, polypeptide 1 [CYP2E1]), and (4) a gene whose expression is under the control of an inducible promoter in an experimental situation (placing a gene adjacent to the metallothionine promoter allows induction by treatment with metal ions, e.g., Cd^{2+}).

Resequencing: Repeating the sequence determination of a DNA fragment.

Restriction enzymes: Enzymes that constitute primitive defense mechanisms of bacteria. Such enzymes recognize specific DNA sequences (usually 4–8 nucleotides in length). Upon binding to the recognition sequence, nuclease activity cleaves the DNA strands.

Restriction fragment length: Restriction enzymes recognize specific sequences in polymorphisms (restriction fragment length polymorphism [RFLP]) to which they can bind and cause cleavage of the DNA. Therefore, these enzymes can be used to perform relatively crude interrogation of a DNA sequence. An enzyme recognizes a particular sequence of DNA bases and will cleave a segment of DNA at each position that this sequence appears. The DNA segment is cut into fragments that can be visualized by agarose gel electrophoresis. In a DNA mixture containing different alleles, disruption of a restriction enzyme recognition site due to the nucleotide polymorphism results in variation in the size of the restriction fragments.

Ribonucleic acid (RNA): The nucleic acid polymers messenger RNA (mRNA) and transfer RNA (tRNA) translate and transcribe the DNA code into peptides and proteins. RNA also functions as both regulatory molecules (rRNA) and as the genetic code in certain viruses.

Sensitivity: A measure of clinical validity: the proportion of truly diseased persons identified as diseased by a positive test result (true positives/(true positives + false negatives)) [Last 1988]. For analytical validity, sensitivity indicates how good the test is at identifying the marker or agent when present.

Sequencing: The determination of the order of the nucleotides (chemical bases) in a DNA or RNA molecule or the order of the amino acids in a protein [DOE 2001].

Single nucleotide polymorphism: A polymorphism resulting from a single base pair difference between alleles.

Sister chromatid exchange: Crossover between sister chromatids. This can occur either in the sister chromatids of a tetrad during meiosis or between sister chromatids of a duplicated somatic chromosome [Jorde et al. 1997].

Sister chromatids: The two identical strands of a duplicated chromosome joined by a single centromere [Jorde et al. 1997].

Specificity: A measure of clinical validity: the proportion of truly nondiseased persons identified as nondiseased by a negative test result (true negatives/true negatives + false positives) [Last 1988]. For analytical validity, specificity indicates how good the test is at correctly identifying the absence of the marker or agent.

Stochastic process: A sequence of events governed by probabilistic laws [Karlin and Taylor 1975].

Susceptibility: The way individuals respond to occupational and environmental contaminants. It usually implies an indicator of degree of risk based on exposure and inherited factors.

Technical variability: Variation in results observed in laboratory measurements in the same sample over time.

Toxicogenomics: The study of changes in expression of numerous genes or gene products due to toxicant-induced exposures.

Transcriptomics: The study of transcriptosomes.

Transcriptosome: The full complement of DNA transcripts (RNA molecules) required for the structure and function of an organism.

Transgenic animal: A genetically engineered animal that carries a gene or genes of another species. Typically, a transgenic mouse or rat carries a human disease gene (e.g., cystic fibrosis, Tay-Sachs disease, or one of the genes associated with breast cancer such as *BRCA1*).

Transitional study: A study that uses biomarkers and bridges the gap between laboratory and population-based studies. The goals of transitional studies may include characterization and validation of biomarkers or optimizing the conditions for using the biomarker. Transitional studies differ from etiological research because the biomarker is generally the outcome or dependent variable as opposed to the disease [Schulte and Perera 1997].

Validation: The process by which a test's performance is evaluated both in the laboratory (e.g., accuracy, precision, sensitivity, and specificity) and in the population (positive predictive value, penetrance, prevalence and other environmental factors) to evaluate the clinical utility of the test. Validation ensures that the test produces accurate and reliable data so that sound medical and health-related decisions can be made based on a test's results [Lee et al. 2005].

Xenobiotic: A natural substance that is foreign to the body.

CHAPTER 1

INTRODUCTION

In the past decade, remarkable advances have been seen in the technology of genetic analysis as well as in the rapid expansion of its use, particularly in human studies and clinical medicine [Kelada et al. 2003; Burke et al. 2002; Little et al. 2002]. Today it is possible to produce a complete record of an individual's genetic makeup. Understanding the genetic components of disease and the interaction between genetic and environmental factors has increased, and the application of this knowledge to the workplace raises medical, ethical, legal, and social issues [Clayton 2003; Ward et al. 2002; McCunney 2002; Christiani et al. 2001; Rothstein 2000a; Schulte et al. 1999; Lemmens 1997; Barrett et al. 1997; Van Damme et al. 1995; Gochfeld 1998; Omenn 1982]. The scientific impact of genetic information can be great, but that impact can be intertwined with other contentious workplace issues. Some of the medical, ethical, legal, and social issues need to be addressed soon because genetic information is already being used in the workplace [Khoury 2002; AMA 2004, 1999]], and legislation has been recently enacted in the U.S. [U.S. Congress 2008]. Debate continues about the use of genetic information and the access to this information by employers, potential employers, insurers, and relatives of the workers being tested, and even about an individual's rights to receive his or her own test results [Renegar et al. 2006; Mitchell 2002]. In some state statutes, genetic information is defined broadly. The laws extend beyond the results of deoxyribonucleic acid (DNA) analysis to encompass family health histories and the results of common laboratory tests for gene products, such as messenger ribonucleic acid (mRNA) or proteins [Kulynych and Korn 2002]. Various professional organizations and authoritative groups have begun to address the issues of genetics in the workplace [Genetics and Public Policy Center 2006; ACOEM 2005; ASCO 2003; CDC 2003; Goel 2001; NBAC 1999].

The purpose of this report is to consolidate the diverse literature and opinions on genetics in the workplace, to flag important issues, and to provide some considerations for current and future research and practice. Occupational safety and health professionals have a growing interest in understanding gene-environment interactions at the mechanistic and population levels that may contribute to prevention and control strategies. The framework for considering the contribution of genetics to occupational disease and injury can be seen in the exposure-disease paradigm that arose from the efforts of the National Academy of Sciences (NAS) in the 1980s [NRC 1987]. This framework (Figure 1–1) identified biomarkers to assess exposure, effects of exposure including disease, and susceptibility. Genetic biomarkers generally pertain to susceptibility. Biomarkers can complement traditional tools such as health and work history questionnaires, exposure measurements, death certificates, and job exposure matrices.

Biomarkers of exposure comprise several of the steps in the paradigm. Internal dose is that amount of agent that enters the body, while the biologically effective dose is the amount of agent that reaches the target site. For example, if an agent causes liver toxicity, then the biologically effective dose is

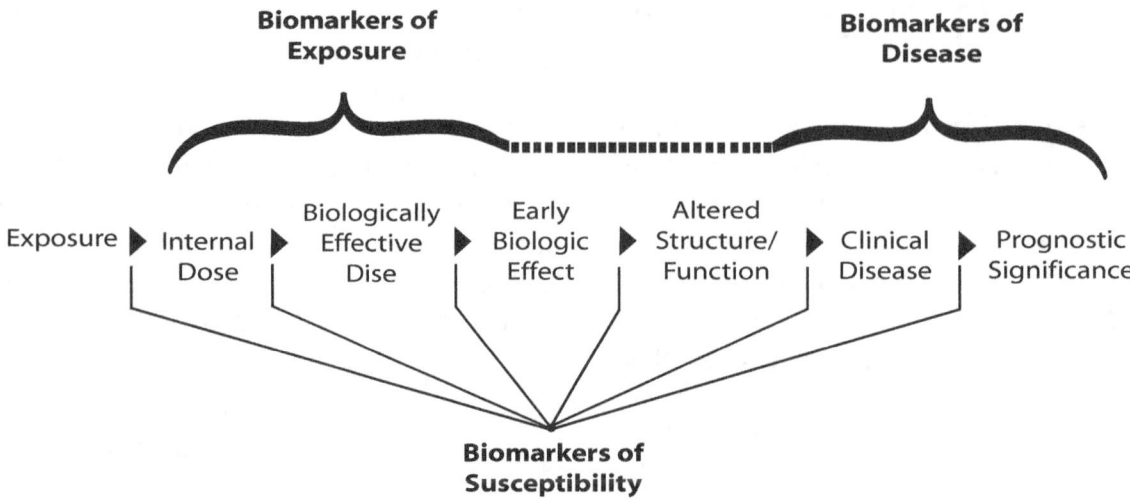

Figure 1-1. Continuum from exposure to disease. (Adapted from NRC [1987]; Schulte and Perera [1993].)

the amount of agent that reaches the liver. Early biologic effects may still be reversible, and some effects may or may not be on the pathway leading to disease. Early biologic effects have been used as both exposure and disease biomarkers, bridging these two aspects. Altered structure or function describes the state before the disease is diagnosed, but during which changes in the biological activity or structure may be present that could indicate early stage disease.

Susceptibility markers can be indicative of acquired or inherited susceptibility, but for the most part, the term has been used for inherited gene variants that may modify the effect of exposure and, therefore, the resulting consequences. Gene variants can modify various steps in the continuum from exposure to disease [Cherry et al. 2002; McCanlies et al. 2002; Weston et al. 2002; Yong et al. 2001; Tamburro and Wong 1993; Kalsheker and Morgan 1990].

Genetic components can be found in all biomarker categories. For biomarkers of exposure, there is a rich history of research on xenobiotics binding with DNA or protein (called DNA or protein adducts) that indicate exposure [Groopman and Kensler 1999; Gledhill and Mauro 1991; Perera and Weinstein 1982; Ehrenberg et al. 1974]. Biomarkers indicating the effects of exposure range from somatic mutations to disease markers manifesting in chromosomal alterations to changes in gene expression [Boffetta et al. 2007; Rossner et al. 2005; Bonassi et al. 2004, 2000, 1995; Hagmar et al. 2004, 1998, 1994; Liou et al. 1999; Albertini and Hayes 1997].

It is possible to consider genetic research and its applications in terms of a matrix (Figure 1–2). The x-axis of the matrix describes where in the continuum between exposure and disease the investigation is focused. The y-axis shows the types of research or applications that could be conducted. These can range from assay development to mechanistic studies to discovery of gene function through etiological research. Increasingly, etiological research is being conducted using whole genomes as in genome-wide association studies. Transitional studies bridge laboratory and population-based studies by characterizing and optimizing biomarker measurements. The matrix, itself, is filled with either specific genes or the genetic biomarkers of interest for the type of study and where in the continuum the research interest lies. Once research is far enough along, applications such as interventions, risk assessments, and monitoring clinical trials can be considered.

Type of research or application	Continuum from exposures to disease						
	Exposure	▶ Internal dose	▶ Biologically effective dose	▶ Early biologic effect	▶ Altered structure/ function	▶ Clinical disease	▶ Prognostic significance
Gene discovery and assay development							
Mechanistic studies							
Transitional studies							
Etiological studies							
Applications: risk assessments, interventions, clinical trials, screening, monitoring, legal proceedings							

Figure 1–2. Matrix of possible uses of susceptibility biomarkers. Note: The horizontal (x) axis indicates a continuum of biomarkers between exposure and resultant effects. At each arrow, a susceptibility biomarker can be considered to intervene.

The work on chronic beryllium disease (CBD) is a good example of the contribution of genetics to occupational safety and health as it covers the gamut of basic research using animal models, to population studies, to the development of policy. A transgenic animal is currently being evaluated for a phenotype with characteristics of CBD. The animal model would be useful to develop and evaluate intervention strategies. In addition, research was done to assess the prevalence of genetic risk factors in workers exposed to beryllium [McCanlies et al. 2007; Tinkle et al. 2003; Weston et al. 2002]. Tests to assess the risk due to major histocompatibility complex, class II, DP beta 1 (HLA-DPB1) polymorphisms have low positive predictive value (PPV) [Weston et al. 2002]. However, as research findings complete the matrix, this should provide valuable information about the development of an exposure recommendation that would protect all workers.

CHAPTER 2

THE ROLE OF GENETIC INFORMATION IN OCCUPATIONAL DISEASE

Genetic information can reveal whether a change occurred in a person's genetic material (e.g., a change in one's DNA, RNA, etc.; also known as acquired genetic effects) as a result of exposure to a harmful agent. Genetic information can also indicate inherited characteristics, such as a gene that interacts with environmental agents to increase or decrease risk of disease. A distinction exists between genetic tests designed to detect genes and those that are designed to find changes in genetic materials [Van Damme and Casteleyn 2003; Schulte and DeBord 2000; OTA 1990]. Both kinds of genetic information—inherited characteristics and changes in genetic material—will be discussed in this chapter.

2.1 Gene-Environment Interactions in Occupational Safety and Health Research

In the past, genetic information was rarely considered in epidemiological studies of occupational diseases, largely because there were no tools for precise measurement of genetic differences that might influence exposure-disease relationships in subsets of the population. Historically, occupational chemical exposures were so high that reasonably valid studies of exposure-disease relationships could be performed even if they did not account for genetic variation [Schulte 1987]. However technology and information have progressed so that the relative influence of genetic factors on exposure-disease relationships is relevant as variables in study design and analysis.

Interest in the role of genetic variants has emerged as a result of studies that have demonstrated variability in response to occupational exposures [Yucesoy and Luster 2007; Godderis et al. 2004; Kline et al. 2004; Thier et al. 2003]. The term "response variability" has been used to describe the differences in the type or magnitude of the biological effect that is due to intrinsic or acquired differences between individuals under identical exposure conditions. Various factors contribute to response variability from workplace or environment exposures (see Figure 2–1) [Hattis and Swedis 2001; Grassman et al. 1998].

One factor that contributes to that observed difference in response variability is individual differences in the uptake of agents. Environmental monitoring may indicate identical exposure conditions, but what is actually absorbed into the body from that exposure may differ between individuals. Individual differences result from a range of factors that influence exposure uptake. Biological variability is one such factor. Biological variability can be further subdivided into interindividual variability, the difference between individuals, and intraindividual variability, the difference within an individual over time.

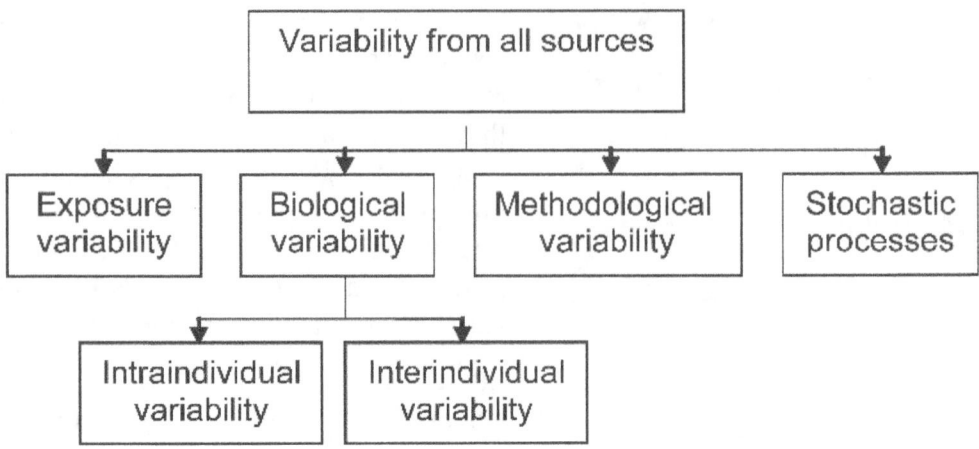

Figure 2–1. Components of variability. (Grassman et al. [1998].)

Genetic polymorphisms contribute to biologic variability and hence may result in interindividual variability in the uptake of agents. For this reason, gene variants are important to study when trying to explain response variability. Our knowledge of the role of gene-environment interactions in occupational diseases has increased in the past few years. Mechanistic studies have focused on the role of specific genes in the development of disease. The majority of studies have investigated carcinogen exposure and polymorphisms in the alleles of genes that code for enzymes involved in xenobiotic metabolism or biotransformation. Metabolism and transformation are intended to remove compounds from the body, but the process may result in the formation of toxic metabolites. Differences in DNA coding result in biological variability in enzymes, which ultimately affect the biotransformation process. Mechanistic studies have consistently reinforced the hypothesis that the biologic variability ultimately affects disease risk by modifying the levels of toxic metabolites.

Table 2–1 lists research involving genetic polymorphisms and occupational/environmental exposure. The trend in research today is to evaluate multiple genes in a study rather than assess the role of single genes. In addition to carcinogenesis, there has been substantial focus on respiratory diseases and allergic responses [McCanlies et al. 2003; Kalsheker and Morgan 1990; Brain et al. 1988; Rystedt 1985; Chan-Yeung et al. 1978]. Beyond that, studies on the gene-occupational environment interaction for such conditions as diabetes, myocardial infarction, and immune function defects have been documented [Omori et al. 2002; Nakayama et al. 2002; Oizumi et al. 2001; Spiridonova et al. 2001; Altshuler et al. 2000; Keavney et al. 2000; Fujisawa et al. 1996; Walston et al. 1995].

According to the Human Genome Project (HGP), which focused on the similarities of the human genome, the differences in the genome should be around for 0.1% [HGP 2009]. After publication of the human genome maps, greater variation was observed than expected. The National Institutes of Health (NIH) initiated the International HapMap Project in 2002. HapMap focused on single nucleotide polymorphisms (SNPs) and their phenotypic variation and relationship to disease. However, other variation exists in the genome structure including deletions, duplications, inversions and copy number variants (CNV). Redon et al. [2006] identified greater than 10% variation in the genome due to CNV after analyzing DNA from the HapMap Project. Others have reported a 0.4% variation in the genome of unrelated people with respect to CNV [Kidd et al. 2008]. Since CNV can be altered with environmental exposures, the differences in reported variation in CNV may be due to somatic changes in the genome.

Genome-wide association studies have been used to identify specific points of variation (SNPs and CNV) in human DNA. By variation, it is possible to investigate differences between people with a disease and those who are disease-free, thereby determining genetic factors that might be involved in specific diseases. Along with HapMap, databases exist on the Web that archive results from genome-wide association studies [Brookes et al. 2009; Wang WY et al. 2005; Becker 2004; IHC 2003].

A genetic power calculator has been developed to aid researchers in determining relationships between disease and genetic factors [Purcell et al. 2003]. In 2007, the National Institute of Environmental Health Sciences (NIEHS) began the Genes, Environment and Health Initiative (GEI) [GEI 2007]. One part of the program is the Genetics Program, which includes a pipeline for analyzing genetic variation in populations with specific diseases. These kinds of tools and studies help scientists understand the complex interrelationships of diseases and the role of multiple genes in disease processes.

While mechanistic studies have established the mechanistic plausibility that polymorphisms affect disease risk, detecting such an effect in health research has been challenging. Over the last decade, many studies of gene-environment interactions have been reported. Most of these studies did not find an association between genetic polymorphisms and disease risk, or the results could not be replicated. In general, inadequate statistical power makes it difficult to detect an effect or association in occupational or environmental health research. This difficulty is compounded in gene-environment studies that require additional attention to study design due to the increased number of variables under study. Factors related to study design may have made detection of an association difficult in early studies. Researchers have attempted to address the problem of insufficient power in the cancer area by developing larger studies and combining studies [IARC 1997; Rothman 1995]. Genome-wide association studies using recently developed

Table 2–1. Research assessing the role of genetic polymorphisms in occupational disease: selected examples

Genetic Factors	Exposure	Disease/Injury	References
NAT2	Aromatic amines	Bladder cancer	Ma et al. [2004]; Cartwright et al. [1982]
NAT2	Smoking Benzidine	Bladder cancer	Marcus et al. [2000a,b] Carreón et al. [2006b]
NAT1, NAT2, GSTM1, GSTT1, GSTP1, UGT2B7	Benzidine	Bladder cancer	Carreón et al. [2006a]
GSTM1	Smoking	Lung cancer	Sobti et al. [2004]
GSTT1, GSTM1, NAT2	Smoking	Bladder cancer	Hung et al. [2004b]
GSTT1	Ethylene oxide	Bladder cancer	Yong et al. [2001]
GSTM1, NAT2, NF2	Asbestos	Mesothelioma	Hirvonen et al. [1995]; Pylkkanen et al. [2002]
XRCC1, XRCC3, XPD, OGG1	Asbestos	Mesothelioma	Dianzani et al. [2006]
CYP2E1, NQO1	Benzene	Hematotoxicity	Nebert et al. [2002]

Table 2–1 (Continued). Research assessing the role of genetic polymorphisms in occupational disease: selected examples

Genetic Factors	Exposure	Disease/Injury	References
CYP2E1, MPO, NQO1, GSTT1, GSTM1	Benzene	Chronic benzene poisoning	Chen et al. [2007]
OGG1	Smoking (reactive oxygen species)	Lung Cancer	Goode et al. [2002]
XRCC1	Smoking; ionizing radiation	Breast cancer	Goode et al. [2002]
TP53, CCND1, CDKN2A	X-rays	Lung Cancer	Hung et al. [2006]
NF2, XRCC1, XRCC3, XRCC5, ERCC2/XPD, Ki-ras, cyclin D1, PTEN, E-cadherin, TGFB1, TGFBR2	Ionizing radiation	Meningioma	Sadetzki et al. [2005]
ALAD	Lead	Lead toxicity (nervous, hematologic, renal, reproductive systems)	Kelada et al. [2001]
HLA-DPB1	Beryllium	Chronic beryllium disease	McCanlies et al. [2003]
MCP-1	Amyloid	Carpal tunnel syndrome	Omori et al. [2002]

Table 2–1 (Continued). Research assessing the role of genetic polymorphisms in occupational disease: selected examples

Genetic Factors	Exposure	Disease/Injury	References
PPAR-γ Calapain, PIR6.2	Impaired glucose tolerance	Diabetes	Altshuler et al. [2000] Stumvoll et al. [2005]
ADRB3	Body mass index (BMI)	Diabetes	Oizumi et al. [2001]; Fujisawa et al. [1996]; Walston et al. [1995]
ACE	Total cholesterol, low-density lipoprotein cholesterol, and BMI	Hypertension; myocardial infarction	Spiridonova et al [2001]; Keavney et al. [2000]
IL-3, IL-4R	Allergens	Asthma	Cookson [2002]
IL-4	Human immunodeficiency virus (HIV)	HIV infection (immune function defects)	Nakayama et al. [2002]
PON1	Organophosphate pesticides	Acute toxicity; respiratory effects	Battuello et al. [2004]; Cherry et al. [2002]
TNF-α	Silica	Silicosis	McCanlies et al. [2002]
BRCA2, AR, CYP17	Electromagnetic fields	Male breast cancer	Weiss et al. [2005]

Table 2–1 (Continued). Research assessing the role of genetic polymorphisms in occupational disease: selected examples

Genetic Factors	Exposure	Disease/Injury	References
BRCA2, AR, CYP17	Electromagnetic fields	Male breast cancer	Weiss et al. [2005]
UGT2B7	Benzidine	Bladder cancer	Lin et al. [2005]
MPO, MnSOD	Smoking, PAH	Bladder cancer	Hung et al. [2004a]
CYP2D6	Pesticides	Parkinson's disease	Elbaz et al. [2004]
MnSOD, NQO1	Pesticides	Parkinson's disease	Fong et al. [2007]
SP-B	Chromate	Lung cancer	Ewis et al. [2006]

large scale SNP platforms are capable of discovering loci associated with relative risks too modest to detect through smaller studies [Hunter et al. 2007]. However, genome-wide association studies with regard to common variants and disease have only moderate predictive power and collectively only explain a small fraction of the genetic component of a disease [Goldstein 2009; Kraft and Hunter 2009]. The true impact of these types of studies may lie in their ability to identify new pathways of underlying diseases [Hirschhorn 2009]. Several aspects are still not resolved in the study of gene-environment interactions in the occupational setting. These include: errors of measurement, testing of multiple hypotheses, interactions, and Mendelian randomization [Vineis 2007].

In the area of cancer research, some examples of polymorphisms of biotransformation enzymes that have been widely studied include glutathione S-transferase M1 (*GSTM1*), glutathione S-transferase theta 1 (*GSTT1*), cytochrome P450, family 1, subfamily A, (aromatic compound-inducible), polypeptide 1 (*CYP1A1*), N-acetyltransferase 2 (arylamine N-acetyltransferase) (*NAT2*), and nicotinamide adenine dinucleotide phosphate (NADPH) dehydrogenase, quinone 1 (*NQO1*). Consistently, the *GSTM1*[null] genotype has been shown to be a risk factor for tobacco-related lung cancer [Olshan et al. 2000]. This association, like that of the association between *GSTM1*[null] and esophageal dysplasia [Roth et al. 2000], gives clues to the etiology of disease. Similarly, *NAT2* is associated with arylamine-related bladder cancer [Marcus et al. 2000a], and *NQO1* has been shown to be important for benzene-associated leukemia [Rothman et al. 1997]. However, the contribution of these biotransformation genes to the overall risk of disease development is small.

One example that illustrates the relative effects of genes, environmental exposures, and their interaction is the role of the pro-inflammatory cytokine tumor necrosis factor (*TNF*-α) in silicosis. The *TNF*-α-*308* polymorphism modifies the risk of silicosis among silica-exposed workers [McCanlies et al. 2002; Yucesoy et al. 2001]. Two methods were used to determine the extent to which genetics contributes to the risk of disease beyond silica exposure. The predicted probabilities of disease were calculated for each individual and then ranked and categorized into exposure deciles. Individuals in the highest exposure decile were at fourfold increased risk of developing silicosis compared with those in the lowest decile. However, when the comparison was restricted by genotype, individuals in the highest exposure decile who had *TNF*-α-*308* were at eightfold risk. Using the second method, the investigators employed relative operating characteristic (ROC) curves to investigate the effect of genetic information on predicting whether an individual would have disease. No difference was seen when the curve generated using exposure alone was compared with that generated using both exposure and genetic data. The overall conclusion from this study was that while genotype plays a role in characterizing risk groups and disease mechanisms, the genotype is unlikely to be a predictor of disease for an individual [McCanlies et al. 2002].

2.2 The Effect of Occupational and Environmental Exposures on Genetic Material

Damage to DNA or other hereditary material of somatic cells can be used to evaluate exposures and, potentially, disease risk. A variety of genetic biomarkers has been used to show exposure or effects from environmental or occupational exposures [Abdel-Rahman et al. 2005; Holeckova et al. 2004; Godschalk et al. 2003; Albertini et al. 2003; Medeiros et al. 2003; Toraason et al. 2001; Ewis et al. 2001; Lu et al. 2001; Wu et al. 2000]. An increasing number of studies has evaluated gene-environment interactions, but much more of the literature describes the effect of occupational exposures on genes and other genetic material. The theoretical underpinnings of this research have grown out of the assessment of workers and populations exposed to radiation from nuclear weapons and nuclear medicine techniques [Albertini 2001; Moore and Tucker 1999]. Somatic mutations, DNA adducts and protein adducts, and cytogenetic changes have frequently been used as biological measures of exposure and, in some cases, as biomarkers of effect. Radiation studies suggest a strong linear dose response correlation between exposure and observed mutation frequency. The evaluation of changes in genetic material is usually part of research studies that investigate the effects of exposure or can be part of periodic physical examinations performed specifically for genotoxic agents in the workplace. Table 2–2 presents the types of genetic materials that have been used to assess exposure or effect.

Whether these changes are biomarkers of effect, and ultimately risk factors for disease, depends on the extent to which the association with the disease has been affirmed. While numerous cross-sectional studies have consistently identified cytogenetic changes associated with exposures to genotoxic substances or agents, only longitudinal analysis is best suited to identify which genetic biomarkers are risk factors for disease. For example, using prospective designs, an increase in chromosomal aberrations has been associated with an increased risk of cancer development [Boffetta et al. 2007; Bonassi et al. 2007, 2004, 2000; Rossner et al. 2005; Liou et al. 1999; Hagmar et al. 1998].

Gene activity also can be altered without changing the DNA sequence. Various epigenetic processes including methylation, acetylation, phosphorylation, ubiquitylation, and sumolyation as well as chromatin modification can affect gene activity. A wide variety of illnesses, behaviors, and health indicators have some level of evidence linking them with epigenetic mechanisms, including cancers, cognitive dysfunction, respiratory, cardiovascular, reproductive, autoimmune and neurobehavioral illnesses [Weinhold 2006]. Heavy metals, pesticides, diesel exhaust, tobacco smoke, polycyclic hydrocarbons, hormones, radioactivity, viruses, bacteria, and nutrients are known or suspected to influence epigenetic processes [Weinhold 2006]. While a comprehensive view of epigenetics in relation to occupational disease still has not been developed, a large role and further research is necessary since epigenetics may have a role in understanding occupational and environmental causes of diseases [Wade and Archer 2006].

2.3 Effect of Genes on Biomarker Measurements and Use in Risk Assessments

Increasingly, sophisticated epidemiological studies have been conducted using biomarkers of both susceptibility and exposure, including measurements of effect, to evaluate the exposure levels that cause adverse effects among groups with differing susceptibility. For example, a study among physicians assessed how biomarkers of the biologically effective dose and polymorphisms of the metabolizing genes for glutathione S-transferase (GST) interacted to predict smoking-related risk of lung cancer [Perera et al. 2003]. The study found that, among then current smokers, DNA adduct levels were associated with a threefold risk of lung cancer, after controlling for GST genotype. The $GSTM1^{null}$/glutathione S-transferase pi ($GSTP1$) Val genotype was associated with a fourfold risk of lung cancer overall, especially among former smokers, and did not vary by adjustment for adduct levels. Among people with lung cancer, adduct levels were significantly higher among then current and former smokers with the $GSTM1^{non-null}$/$GSTP1$ Ile genotype, suggesting a complex interaction between genotype and adduct formation with smoking exposure.

Genetic susceptibility biomarkers have been used in risk assessment models to determine the impact of the role of genetic polymorphisms of metabolism genes on risk estimates [El-Masri et al. 1999; Bois et al. 1995]. Susceptibility biomarkers may reflect variation in exposure, kinetics, and effects and are therefore important to consider in risk assessments. To gauge the impact of genetic markers on risk assessment, El–Masri et al. [1999] conducted a simulation study of cancer risk estimates for exposure to dichloromethane. The risk estimates were 23% to 30% higher when an effect-modifying polymorphism ($GSTT1$) was not included in the models. Mechanism-based modeling has the potential to decrease uncertainties across and within species and exposure scenarios and to quantify pathways and complex relationships within gene networks [Toyoshiba et al. 2004].

An example of the utility of assessing the impact of genetic changes was illustrated by the International Agency for Research on Cancer (IARC) when the agency incorporated the alkylation and genotoxic effects of ethylene oxide in considering its carcinogenicity [IARC 1994]. With only limited epidemiological evidence of ethylene oxide carcinogenicity in humans, IARC relied on supporting evidence that demonstrated this chemical causes a dose-dependent increase in macromolecular adducts and other biomarkers that reflect genotoxic damage. This conclusion was supported by a wealth of animal research linking ethylene oxide exposure to cancer.

2.4 Genomic Priorities and Occupational Safety and Health

Advances in molecular genetics have resulted in considerable progress in identifying the genetic basis of diseases; however, the contribution of genetic information to preventing significant occupational health hazards has been limited. Many of these health hazards may not have a strong genetic component. At some point, research priorities will need to be established so that resources are spent wisely on research to evaluate occupational and environmental health hazards in which genetic information will contribute to understanding the etiology, mechanisms, or possible means of control of those hazards [Schulte 2007; Merikangas and Risch 2003].

Table 2–2. Genetic biomarkers of exposure or effect of exposure: selected examples

Genetic biomarkers	Exposures*	References
DNA Adducts	Smoking	Taioli et al. [2007]
	PAH	Peters et al. [2008]
	Diesel exhaust	Artl et al. [2007]
Hemoglobin Adducts (surrogates for DNA adducts)	Acrylamide	Hagmar et al. [2005]
	Styrene	Teixera et al. [2008]
	1,3-Butadiene	Boysen et al. [2007]
DNA Strand Breaks	Roofing asphalt	Toraason et al. [2001]
	Antineoplastic drugs	Marczynski [2006]
		Deng H [2005]
Sister Chromatid Exchanges	Anesthetic gases	Bilban et al. [2005]
	Gasoline	Celi and Akbas [2005]
	Styrene	Teixeira et al. [2004]
Micronuclei	Ionizing radiation	Mrdjanovic et al. [2005]
	Trivalent chromium	Medeiros et al. [2003]
	PAH	Pavanello et al. [2008]
	Styrene	Migliore et al. [2006]
Chromosomal Aberrations	Benzene	Holeckova et al. [2004]
	X-rays	Milacic [2005]
	Fenvalerate	Xia et al. [2004]
Reporter Genes		
HPRT	Styrene	Abdel-Rahman et al. [2005]
	Radiation	Jones et al. [2001]
GPA	Pesticides	Takaro et al. [2004]
Oncogene Mutations	Arsenic	Wen et al. [2008]
	Coal emissions	Keohavong et al. [2005]
Gene Expression	Smoking	Sexton et al. [2008]
	Cisplatin	Gwosdz et al. [2005]
Protein Expression	Benzene	DeCoster et al. [2008]
	Pollution	Joo et al. [2004]

*The exposures shown here are examples of the substances or agents studied and not a complete list.

CHAPTER 3

INCORPORATING GENETICS INTO OCCUPATIONAL HEALTH RESEARCH

A major influence on genetic research in occupational health is the exponential increase in scientific advances brought about by the sequencing of the human genome and advances in molecular biology techniques. One such advancement is the polymerase chain reaction (PCR), which has been heralded as one of the most important scientific techniques in molecular biology because it allows the fast and inexpensive amplification of small amounts of DNA [NHGRI 2004]. Microarray technologies have allowed laboratories to study tens of thousands of genes and their expression products, which have greatly increased the number of genes that can be studied at the same time [Collins et al. 2003]. Restriction fragment length polymorphism (RFLP) uses restriction enzymes to cut a piece of DNA, resulting in varying fragments that can be used to identify gene variants.

Because of these scientific advances, genetics has begun to transform research questions and study designs in the applied sciences of public and occupational health [Shostak 2003]. Genetic research studies provide new ways to study risks by evaluating genetic damage and gene-environment interactions. As discussed in Chapter 2, ongoing research focuses on damage to genetic material, alteration of gene function, and interaction of genes with each other and with occupational/environmental factors that increase or decrease risks. Initiatives by federal and other agencies have been created to further the science and improve communication of results.

In 1998, NIEHS initiated the Environmental Genome Project (EGP). It has dual goals of providing information about how individual genetic differences affect disease from environmental agents, and in response to this information will propose appropriate environmental or public health policies [Dahl 2003]. EGP comprises five major research activities: (1) developing mouse models to determine the functional significance of genetic polymorphisms, (2) conducting resequencing and functional analysis of polymorphic genes that are responsive to environmental insult, (3) developing a GeneSNP database that integrates gene, sequencing information, and polymorphic data into gene models, (4) focusing on ethical, legal, and social issues to help understand the implications of genetic research, and (5) developing a collaborative research program among multidisciplinary groups to plan novel and innovative molecular epidemiology studies of environmental-induced diseases [EGP 2008]. Initiatives such as these can aid researchers and help address issues in genetic research. In addition in 2007, NIEHS launched the Exposure Biology Program (EBP), which will focuses on the development of technologies to measure environmental exposure that interacts with genetic variation to result in human disease [EBP 2007].

The incorporation of genetics into occupational safety and health research generally requires collecting biological specimens from participating workers, analyzing those specimens, and developing test and study results. This is in some ways analogous to administering a test to a patient in a clinical setting; however, the intent is usually different. That is, the use of a genetic test or assay in a research setting is to answer scientific questions and to obtain generalizable knowledge. Validated genetic factors can be used as independent and dependent variables or effect modifiers in animal and human studies. Ultimately, however, some of these research tests will be candidates for use in clinical settings. Hence, genetic research on workers may be seen as existing on a continuum between laboratory development and clinical use, as shown in Figure 3–1 [Schulte 2004; Burke et al. 2002; Khoury 2002].

Issues of interpretation and communication of results are critical in workplace research involving genetics. Two types of results, individual or group, may be generated at different times and may have different meanings. Most results to date in occupational safety and health have shown effects at the group level. Interpretation at the individual level is more complex as a variety of factors interplay with each other to affect risk. In addition, the interpretation of research results with respect to overall risk may be uncertain. The investigators may not know if the test or study results are meaningful or clinically relevant. Many times the purpose of the study is to evaluate the meaning of the test or assay, to learn about population characteristics of a genetic marker, or to optimize the assay.

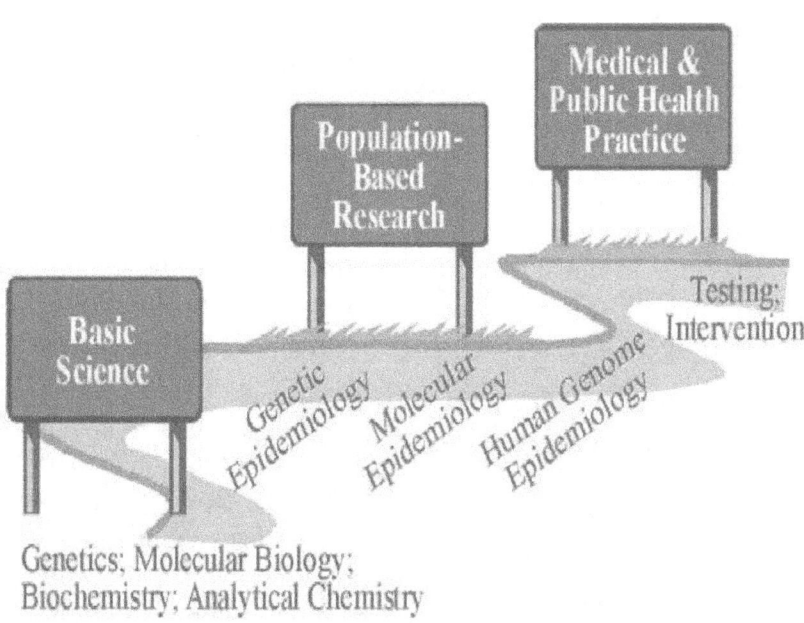

Figure 3–1. Continuum from basic science to medical and public health practice. (Schulte [2004].)

3.1 Validity and Utility Issues of Genetic Assays

Genetic assays or tests may have great value predicting disease risk factors and may establish a new approach in the primary prevention of many chronic diseases. Such tests could allow for the identification and elimination of environmental risk factors that lead to clinical disease among persons with susceptibility genotypes [Khoury and Wagener 1995]. Before such tests are used in practice, the tests need to go through a validation process (see Table 3–1). Validation ensures that the test produces accurate and reliable data so that sound medical or health-related decisions can be made based on this information [Lee et al. 2005].

Researchers may be learning about characteristics of the assays under various conditions and what the assays mean in relation to exposure, disease, susceptibility, or risk. Validation is a process rather than an end state. It is context-specific and pertains to the particular use of a genetic test. It includes analytical validity (whether the test accurately measures the specific genetic property of interest), clinical validity (what the test means in relation to exposure, health, and risk in populations or individuals), and clinical utility (whether the test is actually useful and feasible in clinical or population settings) [Burke et al. 2002].

Validity has been recently addressed by several groups [Constable et al. 2006; HGP 2006; NCI 2006; Barker 2003; Burke et al. 2002; IPCS 2001]. A model process for evaluating data on emerging genetic tests was developed by the U.S. Task Force on Genetic Testing [1998]. More recently, this process has been addressed by a collaborative group sponsored by the Centers for Disease Control and Prevention (CDC) [ACCE 2007; Burke et al. 2002]. This group, called the ACCE core group, takes its name from the four components of evaluation: **a**nalytical validity; **c**linical validity; **c**linical utility; and **e**thical, legal, and social implications and safeguards. The effort builds on a methodology described by Wald and Cuckle [1989] for evaluating screening and diagnostic tests. The ACCE process includes collecting, evaluating, interpreting, and reporting data about DNA (and related) testing for disorders with a genetic component in a format that allows policymakers to have access to up-to-date and reliable information for decision-making. The ACCE model contains a list of 44 questions, targeting the four areas of ACCE, to develop a comprehensive review of a candidate test for potential use [ACCE 2007; Burke et al. 2002]. The International Programme on Chemical Safety (IPCS) discussed validation and the use of genetic markers in risk assessment, and the National Cancer Institute (NCI) developed a list of questions to evaluate whether a genetic test is appropriate for screening purposes [NCI 2006; IPCS 2001].

3.1.1 Analytical Validity

The first step in establishing validation of a test is analytical validity, which focuses on the ability of the test to measure accurately and reliably the marker/genotype of interest. The four components of analytical validity are analytical sensitivity, analytical specificity, laboratory quality control, and test robustness [ACCE 2007; Burke et al.

Table 3–1. Considerations in the Validation of Genetic Tests

Analytical Validity	Clinical Validity	Clinical Utility
Is the test qualitative or quantitative?	How often is the test positive when the condition is present?	What is the natural history of the condition?
How often is the test positive when the marker is present?	How often is the test negative when the condition is not present?	What is the impact of a positive (or negative) test on patient care?
How often is the test negative when the marker is not present?	Are there methods to resolve clinical false positive results in a timely manner?	If applicable, are diagnostic tests available?
Is an internal quality control program defined and externally monitored?	What is the prevalence of the condition in this setting?	Is there an effective remedy, acceptable action, or other measurable benefit?
Have repeated measurements been made on specimens?	Has the test been adequately tested on all populations to which it may be offered?	Is there general access to that remedy or action?
What is the within- and between-laboratory precision?	What are the positive and negative predictive values?	Is the test being offered to a socially vulnerable population?
If appropriate, how is confirmatory testing performed to resolve false positive results in a timely manner?	What are the genotype/phenotype relationships?	What quality assurance measures are in place?
What range of patient specimens have been tested?	What are the genetic, environmental, or other modifiers?	What are the results of pilot trials?
How often does the test fail to give a usable result?	Do well-designed studies exist evaluating the relationship between the marker and the condition?	What health risks can be identified for follow-up testing and/or intervention?
How similar are results obtained in multiple laboratories using the same or different technology?		What are the financial costs associated with testing?
How robust or rugged is the test?		What are the economic benefits associated with actions resulting from testing?
What factors affect the test results?		What facilities/personnel are available or easily put in place?
		What educational materials have been developed and validated, and which of these are available?
		Are there informed consent requirements?
		What methods exist for long-term monitoring?
		What guidelines have been developed for evaluating program performance?
		What is known about stigmatization, discrimination, privacy/confidentiality, and personal/family social issues?
		Are there legal issues regarding consent, ownership of data, and/or samples, patents, licensing, proprietary testing, obligation to disclose, or reporting requirements?
		What safeguards have been described, and are these safeguards in place and effective?

2002]. Analytical sensitivity evaluates how well the test indicates the marker/genotype when it is present. Analytical specificity, on the other hand, evaluates the test to determine how well it identifies true negatives and false positives [Schulte and Perera 1993]. Laboratory quality control evaluates the test procedures to ensure they fall within the specified limits, and robustness measures the variability of the test measurement under different analytical and preanalytical conditions.

3.1.2 Clinical validity

The second step in the validation process is to determine the clinical validity or the ability of the genetic test to detect or predict the associated condition (phenotype). The parameters of clinical validity include clinical sensitivity, clinical specificity, prevalence of the condition, positive and negative predictive value, penetrance, and modifiers of the condition, such as other genes and environmental factors. Clinical sensitivity measures the proportion of the population that will get the condition when the test value is positive, while clinical specificity is the proportion of the population that will not get the condition when the test result is negative. Positive Predictive Value (PPV) is the proportion of individuals who will develop the disease given that they have the marker/genotype, while the negative predictive value is the proportion of individuals who test negative who will not get the disease [Weston et al. 2002]. Penetrance defines the relationship between the genotype and the phenotype, so that the expression of the genotype can be determined and is broadly equivalent to the PPV [Constable et al. 2006]. Prevalence measures the proportion of individuals with the genotype who have or will get the condition. An additional factor in clinical validity is heterogeneity—the same disease might result from the presence of any number of several different gene variants or totally different genes altogether [HGP 2006]. Considerations for clinical validity include the design of the study, size of the population, type of test, and the endpoints measured [Constable et al. 2006]. For validation, data needed to establish the clinical validity of the genetic test must be collected under investigative protocols, the study sample should be drawn from a population that is representative of the population for whom the test is intended, and formal validation needs to be established for each intended use of the test [HGP 2006]. Establishing clinical validity may take a lot of time and resources. Yang et al. [2000] estimated that for many adult onset diseases this process could take decades.

3.1.3 Clinical utility

Clinical utility addresses the elements that need to be considered when evaluating risks and benefits associated with the introduction of the genetic test into routine clinical practice. Three strategies are used to help determine whether a test has clinical utility. They are screening to detect early disease, interventions to decrease the risk of disease, and interventions to improve the quality of life. It is important to know the accuracy of the testing methods, the strength of the correlation with the clinical phenotype and the condition, and the utility of the information [Ginsburg and Haga 2006]. Evaluating clinical utility will require checks to ensure that due consideration is given to the complex array of factors that go into establishing clinical utility such as the predictive value, nature of the condition and associated social burdens, and the safety of the treatment and cost-effectiveness of the treatment. An example of a test that has clinical utility is blood cholesterol, which provides the individual with valuable information that can be used for prevention, treatment, or life planning, regardless of results.

3.1.4 Validation of assays for evaluation of exposure or genetic damage

The first step in establishing the analytical validity of a method to test for genetic damage is characterization of the genetic biomarker of interest. Characteristics such as dose-response, biomarker persistence, interindividual and intraindividual variability, methodological variation, correlation with other markers, and correlation with a critical response are crucial [Schulte and Talaska 1995; Vineis et al. 1993]. The establishment of a laboratory quality assurance program is essential before any assays of genetic damage are used in research or practice. After analytical validity has been established, clinical validity must be determined to demonstrate if the biomarker occurs in the population. Once the analytical and clinical validity are recognized, the risks associated with that biomarker at an individual level can be evaluated.

3.1.5 Validation of assays used to evaluate genetic polymorphisms

Genotyping assay results are subject to a variety of laboratory errors, such as misincorporation rate when using the polymerase chain reaction (PCR), differences in reagents, PCR artifacts, contamination by foreign DNA, and differences in efficiency of allele detection [Millikan 2002]. Differences in interpreting genotyping results and errors contained in on-line genotyping databases are a source of variation. Errors can be found in many on-line databases, as not all sequence information or the order and location of all human genes have been confirmed. Differences in gene frequencies across populations in epidemiological studies can lead to small sample sizes and the subsequent lack of power. Deciding which genes and their variants to evaluate when studying disease-gene association in an epidemiological study can introduce bias because some genes may not be selected that may have a role in the disease or genes are selected that have no role in the disease [Peltonen and McKusick 2001]. Validation of a genetic test for worker populations is difficult, but has been demonstrated.

The PPV of any genetic test is dependent on the prevalence of the disease and genetic trait in the population and the relative risk of disease for those carrying the trait [Khoury et al. 1993, 1985]. This issue has been demonstrated practically for *HLA-DPB1* [Weston et al. 2002], which has been implicated as a susceptibility gene in beryllium hypersensitivity and CBD [McCanlies and Weston 2004; Rossman et al. 2002; Saltini et al. 2001; Wang et al. 2001, 1999;

Richeldi et al. 1997, 1993]. The prevalence of the genetic trait $HLA\text{-}DPB1^{E69}$ for four major racial/ethnic groups in the United States was determined. Carrier frequencies ranged between 33% and 59%. Based on published studies, CBD disease prevalence among beryllium workers is 3% to 5% [Richeldi et al. 1997; Kreiss et al. 1996]. To estimate the PPV, investigators assumed a disease prevalence of 5% and odds ratios of 35 and 3 for CBD associated with inheritance of $HLA\text{-}DPB1^{E69}$ [Weston et al. 2002] and found that, for a genetic trait prevalence of 33% to 59%, the PPV for $HLA\text{-}DPB1^{E69}$ was 8% to 14% when the odds ratio was 35 and 7% to 9% when the odds ratio was 3. The authors also assumed a higher prevalence rate of 15%, as might be seen in some higher-risk jobs, and the PPV for $HLA\text{-}DPB1^{E69}$ was 25% to 43% when the odds ratio was 35 and 21% to 27% when the odds ratio was 3. Thus, for a tenfold difference of an odds ratio, a modest change in PPV was seen [Weston et al. 2002].

To date, many of the epidemiological studies to validate biomarkers of susceptibility have exhibited a high degree of heterogeneity in their results [Wenzlaff et al. 2005; Pavanello and Clonfero 2004; D'Errico et al. 1996]. In a review of four genetically based metabolic polymorphisms involved in the metabolism of several carcinogens, D'Errico et al. [1996] identified a range of methodological features that lead to discordant results. These include a high proportion of studies using prevalent cases, the frequent use of hospital controls, a low response rate, the use of metabolic ratios as variables, and the lack of adequate adjustment for covariates. In addition, such studies had small sample numbers and weak exposure characterizations [Vineis 1992].

3.1.6 Validation of multiple biomarkers

The use of multiple biomarkers has the potential to increase the understanding of exposure, disease, or susceptibility, but creates the challenge to combine the information from individual markers and interpreting the overall combination of markers. This makes validation more difficult to accomplish since the complexity has increased. Therefore, how composite data will be analyzed and interpreted must be considered when a study is designed. Perera et al. [1992] demonstrated the benefits and extra knowledge that can be gained when they used a battery of biomarkers to assess genetic and molecular damage in residents of a polluted area of Poland. In this study, a common genotoxic model was used to provide a molecular link between environmental exposure and genetic alteration relevant to cancer and reproductive risk. Most of the biomarkers (carcinogen-DNA adducts, sister-chromatid exchanges, chromosomal aberrations, and ras oncogene overexpression) were related to levels of exposure to polycyclic aromatic hydrocarbons (PAHs) in the air.

3.2 Challenges of Genomics and Related Research Areas

Technological developments, such as DNA and gene microarrays, and automated work stations capable of extracting, amplifying, hybridizing, and detecting DNA sequences will present a number of benefits and issues in studying genetic and environmental variables [NRC 2007; Christiani et al. 2001]. The benefits include the ability to study large numbers of genes, practically the entire human genome, in one study or experiment and to have access to data banks containing further information on genomic DNA. The primary attendant issue with this technology includes heightened difficulties in analyzing and interpreting for research participants such large amounts of data [Ermolaeva et al. 1998].

The area of genomics or the study of the expression of gene or gene products has grown exponentially and spun off a variety of other related scientific areas such as toxicogenomics, proteomics (changes in cellular protein expression or function), metabonomics (changes in metabolites), and transcriptomics (changes in mRNA). One similarity among all of these fields is that large amounts of data are generated using high-throughput technologies, such as microarrays.

Toxicogenomics is the study of changes in expression of numerous genes or gene products due to toxicant-induced exposures. As our understanding grows, the science of toxicogenomics has become more useful in occupational and environmental health [Koizumi 2004; Tennant 2002; Henry et al. 2002; Nuwaysir et al. 1999]. Initially, toxicogenomics research has used various cell lines to understand the differences and similarities between species so that the toxicity and changes in genes/gene products could be compared between animals and humans. A need exists for standardization of data collection from microarray experiments, optimization of information, and knowledge management to make comparisons between studies easier and more accurate.

Proteomics is the sister technology to genomics [Kennedy 2002]. Genomics provides information about DNA and RNA, whether it is under- or overexpressed. Messenger RNA is translated into proteins, which provide the structural and functional framework for cellular life. By studying proteins, we can see how a cell or organ responds to an insult. Proteomic researchers are also using high-throughput, sensitive technologies that can be used for protein identification and characterization.

Protein arrays along with genomic data can be used to better understand disease processes and mechanisms of toxicity and to develop biomarkers for diagnosis and early detection of diseases [Hanash 2003; Kennedy 2002]. The enthusiasm for this approach has been somewhat dampened because methodological and bioinformatic artifacts have been identified in some of the initial papers that suggested proteomics may be useful for disease diagnosis [Diamandis 2006]. Sample collection and storage conditions may produce protein profiles that overshadow those generated by the disease. Further adding to the complexity, just as most diseases are not caused by single genes, it is likely that a disease will

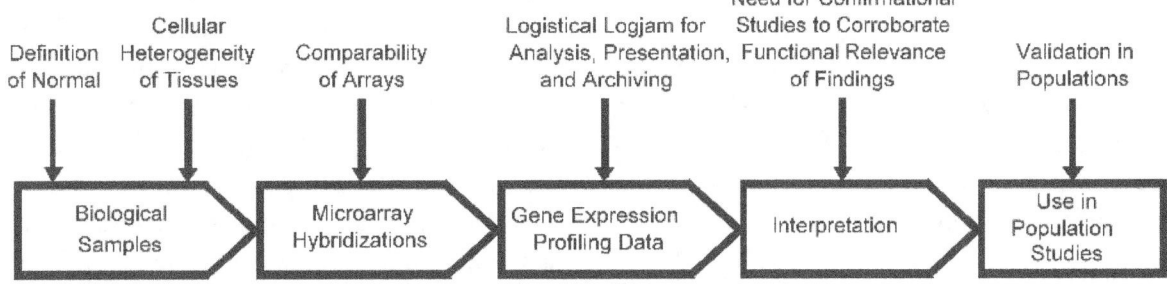

Figure 3–2. Issues in microarray gene expression analysis and use. (Adapted from King and Sinha [2003].)

not be identified by a single protein, but by a protein profile as multiple pathways are involved in disease promotion/progression.

Microarray technologies for DNA and proteins are the leading edge of efforts to use batteries of biomarkers to assess exposure, effect, or susceptibility. Making sense out of complex arrays of genetic expression data and multiple markers is difficult (see Figure 3–2) [King and Sinha 2001]. Much of the literature on the predictive value of diagnostic tests concerns a single test, with brief mention of multiple tests. Similarly, experience to address situations with multiple genes and environmental interactions is lacking. Koizumi [2004] described four ways DNA microarrays could be used in research relevant to occupational health: (1) understanding the mechanistic background of health effects, (2) toxicity testing, (3) search of indicators for hazard prevention and health management, and (4) managing high-risk populations.

New technologies and approaches now allow researchers to focus more on studies of gene-environment interactions that aim to describe how genetic and environmental occupational factors jointly influence the risk of developing disease [Hunter 2005; Christiani et al. 2001]. The study of gene-environment interactions (1) allows for better estimates of population-attributable risk of genetic and occupational factors, (2) strengthens associations between occupational risk factors and disease, (3) provides insight into mechanisms of action, and (4) provides new opportunities for intervention and prevention [Hunter 2005].

The use of DNA microarrays is revolutionizing genetic research. Microarrays have the ability to analyze gene expression patterns, carry out genome-wide mapping, clone members of gene families within and across species, scan for mutations in interesting genes, define genes controlled by particular transcription factors, being used for diagnostic purposes and have application in risk assessments [Beaudet and Belmont 2008; Travis et al. 2003; Blanchard and Hood 1996; Borman 1996]. The expectation of researchers is that the expression of many thousands of genes will be measured before and after an exposure, although in the occupational setting, biomarkers are usually measured after exposure. These technologies will amplify the challenges in handling large amounts of data. Those challenges include data reduction, summation, analysis, and interpretation.

A challenge of array data analysis is that many early studies have shown that different sets of genes are affected by similar exposures. This finding may be due to a lack of standardization of approach in guiding analysis and interpretation. Studies where gene expressions are measured under different conditions need rules for determining what is an outlier (a potentially important deviation in expression). In the context of gene expression data, this is sometimes described as the challenge of detecting a signal, a true difference in expression, amidst the noise of natural variability in gene expression. Whether variability for different genes can be expressed in the same scale of measurement needs to be determined [Wittes and Friedman 1999]. Replication is necessary to determine whether tested genes have the same natural scales of measurement. Even if the same scale of measurement is appropriate for different genes, the question arises as to whether the appropriate method has been used for detecting outliers [Wittes and Friedman 1999]. Challenges attending the interpretation of the deluge of data include identifying genes and their functions, identifying reproducible artifacts, and distinguishing homeostatic from pathologic perturbations.

Different approaches have been suggested to analyze the large amount of gene expression information generated from microarray experiments [IARC 2004]. Statistical methods have been developed and refined for determining which genes, among the thousands on microarray chips, are differentially expressed across experimental conditions or biological states. Adjustments for multiple testing are necessary to balance the risks of type I and type II errors. Multivariate statistical analysis approaches, such as cluster analysis, are often then applied to discover groups of genes that have similar expression patterns on microarrays across different experimental conditions or biological states. These patterns are then examined to explore whether coexpression of the genes provides clues about their regulation or their function.

A need exists for standardization of data collection from microarray experiments, optimization of information, and knowledge management to make data exchange and comparisons between studies easier and more accurate. One approach to standardizing collection, analysis, and dissemination of microarray data was the development of the minimum-information-about-a-microarray-experiment (MIAME) guidelines [Brazma et al. 2001]. MIAME has six essential elements: (1) the study experimental design, or the set of hybridization experiments as a whole, (2) the array design, or how each individual array and each element in the array is used, (3) the sample characteristics (which samples are used and how the samples are extracted, prepared, and labeled), (4) the hybridization procedures and parameters, (5) measurement (description and includes images, quantification, and specifications), and (6) normalization, control types, values, and specifications. This six-part approach has become the standard for some journal submissions.

The utility of microarray data will depend on our ability to interpret and communicate the data. Ultimately, the goal is to integrate this information into risk assessment [Ermolaeva et al. 1998]. In occupational and environmental health, a key decision will be when the technology will be mature enough for regulatory use [NAS 2003]. The NAS, in conjunction with a federal liaison group, identified four challenges to using toxicogenomic microarray data by government agencies: premature use of the data, data interpretation, communications, and information gaps [NAS 2003].

In addition to analyzing a gene for mutations, the need to consider factors affecting gene penetrance and phenotypic expression, such as gene expression and environmental covariates, still exists if genetic information is to be useful for environmental health research or risk assessment.

3.3 The Relationship Between Genetic and Environmental Risk Factors

Genetic factors may modify exposure-effect relationships. That is, the risk of effect or disease attributable to an occupational exposure can be decreased, unchanged, or increased depending on the form of interaction (e.g., additive, multiplicative, or synergistic) between the gene variant and the occupational hazard [Poulter 2001]. This effect modification has both statistical and biological aspects. Statistically, the examination of the joint effects of two or more factors is often discussed and depends on the statistical method (e.g., multiplicative or additive) used to model the interaction. From a biological perspective, effect modification conceptually answers the question of why only one of two similarly exposed individuals develops a disease. The answer, in part, is variability in genetic makeup between individuals [Schulte and Perera 1993].

Gene-environment interactions are the combination of environmental exposures and genetic polymorphisms to bring about an effect. Ottman [1996] described six biologically plausible models relating genotype to exposure (see Figure 3–3). The sixth model showed no interaction between a genetic biomarker and an occupational/environmental risk for the disease. Models A through E in Figure 3–3 describe different scenarios of gene-environment interactions. In Model A, the genetic biomarker or genotype causes an increase in the environmental risk factor, perhaps by increased absorption of the agent into the body. The genotype or genetic biomarker exacerbates the effect of the risk factor in Model B. For example, the genotype results in increased metabolism of a chemical so that greater levels of carcinogenic metabolites are formed. Alternatively, the genotype may result in a change in metabolism so that less carcinogenic metabolites are formed. In Model C, the exposure increases the effect of the genetic biomarker or genotype by causing increased or decreased gene expression. Both exposure and genotype directly increase the risk of disease in Model D. In Model E, the exposure and the genetic biomarker both have the same effect on disease risk, either increasing or decreasing that risk; for example, silica and *TNF-α-308* are both risk factors for silicosis. Each of these models may lead to a different prediction about disease risk in individuals classified by the presence or absence of a high-risk genotype or environmental exposure.

A case study of occupationally induced asthma is given as an example of some of the models. As with the case of silicosis, it appears that in occupationally induced asthma, gene-environment interactions may be important [Barker et al. 2003]. Genes can modify the risk of occupationally induced asthma in two ways. Genes may not have any effect on the disease, but may alter the response to the environmental exposure of the allergen (Model B), or they may be involved in the response to the allergen (Model E).

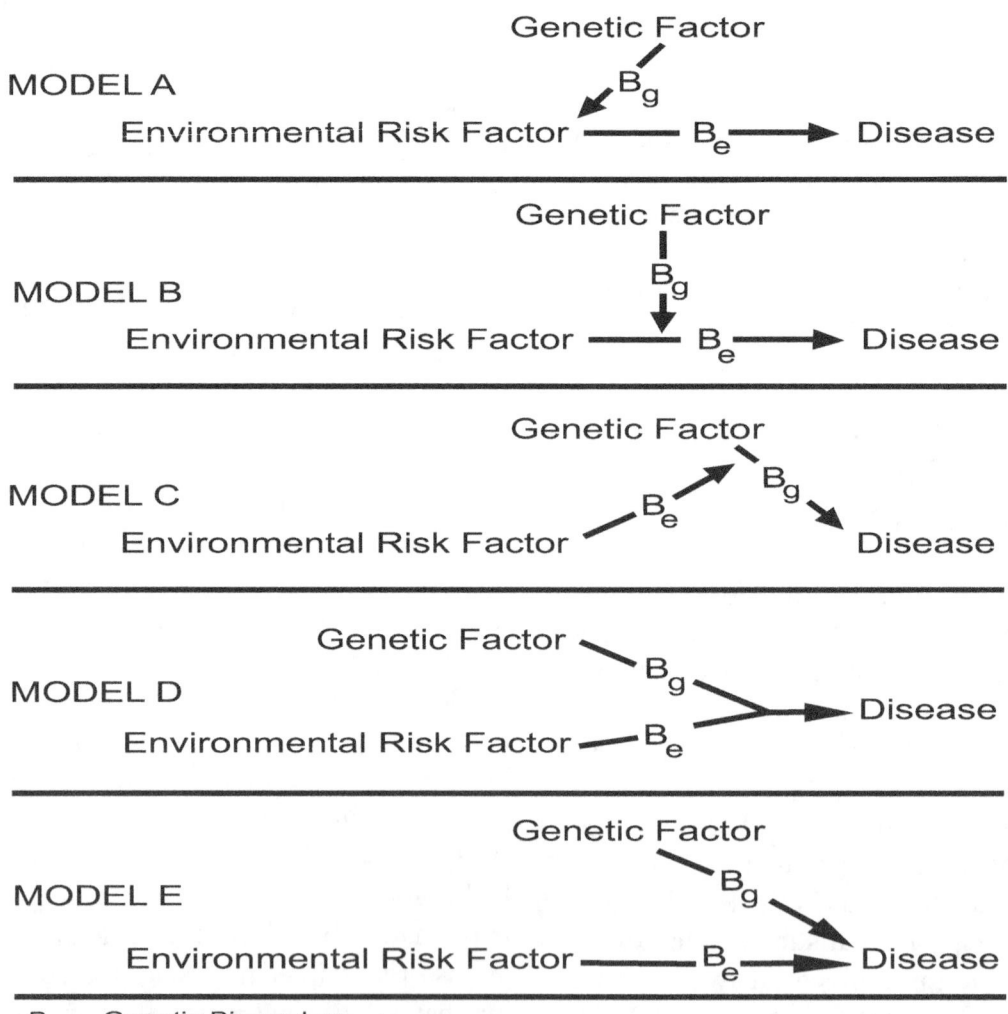

B_g — Genetic Biomarker
B_e — Environmental/Occupational Biomarker

Figure 3—3. Relationships of genetic and environmental risk factors to disease. (Adapted from Ottman [1996].)

Case Study of Occupationally Induced Asthma

Diisocyanates are low-molecular-weight chemicals known to produce occupational asthma. However, only 5% to 10% of those exposed develop disease [Piirila et al. 2001]. An example of how a gene may modulate a response is through biotransformation or metabolism of the xenobiotic. Glutathione S-transferases (GSTs) bind diisocyanates and their metabolites. This binding of the diisocyanates or their metabolites may alter the body's response to the diisocyanate. Two studies of workers with diisocyanate-induced asthma were compared with an exposed, but asymptomatic, control group. Piirila et al. [2001] found that GST polymorphisms altered the response to exposure. $GSTM1^{null}$ (one polymorphism of GST) nearly doubled the risk of developing diisocyanate-induced asthma. In a second study, Mapp et al. [2002] evaluated 92 workers who had toluene diisocyanate (TDI)-induced asthma and compared them with 30 asymptomatic exposed workers. They reported that workers with asthma with more than 10 years of exposure to TDI were less likely to have the *GSTP1* (Val/Val) genotype, a second GST polymorphism. It appears that persons who are homozygous for these alleles have some protection against TDI-induced asthma.

In another example, for different genes, Wikman et al. [2002] investigated genotypes for the N-acetyltransferase-metabolizing enzymes in 109 workers with diisocyanate-induced asthma and 73 asymptomatic workers. The slow acetylator genotype for N-acetyltransferase 1 (arylamine N-acetyltransferase (*NAT1*) conferred a 2.5-fold increase in the risk of developing diisocyanate-induced asthma in general and a 7.8-fold increase in risk for TDI-induced asthma.

In the above example, genes modified the metabolism of the chemical exposure, which increased or decreased the risk of asthma. Genes can also modify the body's immune response to the allergen in occupationally induced asthma. Human lymphocyte antigen (HLA) class 2 molecules have a crucial role in the immune response that occurs in occupational asthma. In one study, 67 workers with TDI-induced asthma and 27 asymptomatic controls were genotyped for three HLA class 2 genes: *DQA1, DQB1*, and *DRB1* [Mapp et al. 2000]. Asthmatics were found to have a significantly higher frequency of specific alleles for *DQA1* and *DQB1*, while controls had higher frequencies of two other alleles for *DQA1* and *DQB1*. Taylor [2001] reported that the HLA class 2 gene, *DR3*, was more prevalent in occupational asthma cases induced by acid anhydrides, suggesting that there may be a contribution of HLA class 2 molecules in individual susceptibility to sensitization and asthma induction.

Interactions between independent causal factors are inevitably affected by exposure response and latency relationships. The term "exposure response" refers to a response or effect seen at a given exposure level. The lower the exposure, the less the response or risk of disease and vice versa. Latency is the period of time between exposure and the onset of the effect or disease. Failure to model dose-response and latency adequately can lead to bias in interaction estimates [Greenland 1993]. In addition, measurement errors, even if independent and nondifferential, can distort interaction assessment. Since both genetic and environmental factors contribute to the etiology of most diseases, each may be expected to modify the effect of each other [Morgenstern and Thomas 1993].

3.4 Analytical Epidemiological Research

Attention to the use of appropriate analytical epidemiological methods is critical for understanding the role of genetics in occupational disease. While many early studies were innovative, they were of limited value because issues of analytical validity of genotyping, possible selection bias, confounding, possible gene-environment and gene-gene interactions, and statistical interaction were inadequately addressed [Little et al. 2002]. Studies with conflicting findings have been common; thus, the role of genetic factors in occupational disease and disorders is not as clear as it might be. Genetic associations with disease or as effect modifiers in exposure response studies suffer from a lack of a systematic approach.

Two CDC efforts have been initiated to address the problem. The first effort is the creation of the Human Genome Epidemiology Network (HuGENet). Activities in this effort are to develop reviews of gene-disease, gene-gene, and gene-environment interactions systematically. Human Genome Epidemiology (HuGE) reviews were established as a means of incorporating evidence from human genome epidemiological studies, i.e., population-based studies to determine the impact of human genetic variation on health and disease [Khoury and Little 2000]. These reviews are systematic, peer-reviewed synopses of the epidemiological aspects of human genes, including prevalence of allelic variants in different populations, population-based information about disease risk, evidence for gene-environment interaction, and quantitative data on genetic tests and services (see Table 3–2) and are carried out according to specified guidelines [Khoury and Little 2000].

The other approach was a result of the CDC-National Institutes of Health (CDC–NIH) HuGE Workshop held in January 2001, which developed a standardized consensus approach for reporting, appraising, and integrating data on genotype prevalence and gene-disease associations [Little et al. 2002]. A checklist (Table 3–3) developed during the workshop was intended to guide investigators in the preparation of manuscripts, to guide those who need to appraise manuscripts and published papers, and to be useful to journal editors and readers. The checklist was not meant to be exhaustive. For use in assessing the role of genetics in occupational disease, the list would need to be extended to specify the approach and as-

Table 3–2. Selected HuGE reviews [HuGENet 2009]

Genes	Diseases	Authors
GST	Colorectal cancer	Cotton et al. [2000]
NAT1 and NAT2	Colorectal cancer	Brockton et al. [2000]
MTHFR	Congenital anomalies	Botto and Yang [2000]
HLA-DQ	Diabetes	Dorman and Bunker [2000]
GSTM1 and GSTT1	Head and neck cancer	Geisler and Olshan [2001]
ALAD	Lead toxicity	Kelada et al. [2001]
NQO1	Benzene toxicity	Nebert et al. [2002]
GSTM1	Bladder cancer Lung Cancer	Engel et al. [2002] Carlsten et al. [2008]
GST	Ovarian cancer	Coughlin and Hall [2002]
GJB2	Hearing loss	Kenneson et al. [2002]
hMLH1 and hMSH2	Colorectal cancer	Mitchell et al. [2002]
APOε	Cardiovascular disease	Eichner et al. [2002]
AR	Prostate cancer	Nelson and Witte [2002]
HLA-DPB1	Chronic beryllium disease	McCanlies et al. [2003]
MTHFR	Leukemia	Robien and Ulrich [2003]
CYP3A4	Breast and prostate cancer	Keshava et al. [2004]
CTSD	Alzheimer's disease	Ntais et al. [2004]
CYP17	Hormone levels	Sharp et al. [2004]
ADH	Head and neck cancer	Brennan et al. [2004]
AR, PGR	Ovarian cancer	Modugno [2004]
MTHFR, MTR	Colorectal neoplasia	Sharp and Little [2004]
ERα and ERβ	Osteoporosis	Gennari et al. [2005]
ERCC2	Lung cancer	Benhamou and Sarasin [2005]
CYP1B1	Breast cancer	Paracchini et al. [2007]
GSTP1	Bladder cancer Lung Cancer	Kellen et al. [2007] Cote et al. [2009]
MPO G-463A	Cancer	Taioli et al. [2007]
NAT1 and NAT2	Bladder cancer	Sanderson et al. [2007]
GST	Hepatocellular cancer	White et al. [2008]
E-cadherin	Cancer	Wang et al. [2008]

Adapted from Burke et al. [2002] and NCI [2006].

Table 3–3. Proposed checklist for reporting and appraising studies of (i) genotype prevalence, (ii) gene-disease associations, and (iii) gene-environment interactions [Little 2004]

	Genotype prevalence	Gene-disease associations	Genotype- environment interaction
1. Purpose of study	Yes†	Detect associations or estimate magnitude of association	Describe joint effects; test specific hypotheses about interaction
2. Analytical validity of genotyping			
Types of samples used	Yes	For cases and for controls	For cases and for controls
Timing of sample collection and analysis, by study group*	e.g., ethnic group	e.g., cases vs. controls	e.g., cases vs. controls
Success rate in extracting DNA, by study group*	e.g., ethnic group	e.g., cases vs. controls	e.g., cases vs. controls
Definition of the genotype(s) investigated; when there are multiple alleles, those tested for should be specified	Yes	Yes	Yes
Genotyping method used (reference; for PCR methods – primer sequences*, thermocyle profile*, number of cycles*)	Yes	Yes	Yes
Percentage of potentially eligible participants for whom valid genotypic data were obtained, by study group	e.g., ethnic group	e.g., cases vs. controls	e.g., cases vs. controls
If pooling was used, strategy for pooling of specimens from cases and controls		Yes	
Quality control measures*	Yes	Including blinding of laboratory staff	Including blinding of laboratory staff
Samples from each group of participants compared (e.g., cases and controls) included in each batch analyzed*		Yes	

Table 3–3. Proposed checklist for reporting and appraising studies of (i) genotype prevalence, (ii) gene-disease associations, and (iii) gene-environment interactions [Little 2004]

	Genotype prevalence	Gene-disease associations	Genotype- environment interaction
3. Assessment of exposures			
Reproducibility and validity of exposure documented			Yes
Categories or exposure scale justified			Yes
4. Selection of study participants			
Geographical area from which participants were recruited	Yes	Yes	Yes
The recruitment period	Yes	Yes	Yes
Recruitment methods for participants whose genotypes were determined, such as random population-based sampling, blood donors, hospitalized participants with reasons for hospitalization	Yes		
Definition of cases and method of ascertainment		Yes	Yes
Number of cases recruited from families and methods used to account for related participants		Yes	Yes
Recruitment rates	Where possible by sex, age, and ethnic group	For cases and controls	For cases and controls
Mean age (±SD) or age range of study participants, and the distribution by sex	Yes	For cases and controls	For cases and controls
Ethnic group of study participants	Yes		
Similarity of sociodemographic (or other) characteristics of participants for whom valid genotypic data were obtained with characteristics of participants for whom such data were not obtained*		Yes	Yes

Implications for Occupational Safety and Health

Table 3–3. Proposed checklist for reporting and appraising studies of (i) genotype prevalence, (ii) gene-disease associations, and (iii) gene-environment interactions [Little 2004]

	Genotype prevalence	Gene-disease associations	Genotype-environment interaction
Steps taken to ensure that controls are noncases*			Yes
5. Confounding, including population stratification			
Design		Yes	Yes
If other than a case-family control design, matching for ethnicity, or adjustment for ethnicity in analysis		Yes	Yes
6. Statistical issues			
Distinguish clearly *a priori* hypotheses and hypotheses generated		Yes	Yes
If haplotypes used, specify how these were constructed	Yes	Yes	Yes
Number of participants included in the analysis, by cell numbers where possible	Yes	Yes	Yes
Method of analysis, with reference, and software used to do this		Yes	Yes
Confidence intervals	Of genotype frequency	Of measures of association with the genotype	
For interaction analysis, 2xK presentation used, or choice of stratified analysis justified			Yes
For interaction analysis, P value for interaction calculated and choice of Wald test or likelihood ratio test specified and justified			Yes
For interaction analysis, null interactions listed			Yes
Assessment of goodness of fit of the model used*		Yes	Yes

†"Yes" indicates essential aspects for reporting; text indicates inclusion with caveats.
*Additional information recorded (ideally in Web-based methods register), but not necessarily presented in journal article.

sumptions for assessing gene-environment interactions and to specify quality characteristics in the measurement of occupational and environmental exposures.

An international effort is underway to help understand the role of genomics in disease and to help solve the methodological problems that have been identified. This effort is called the "network of networks" [Seminara et al. 2007; Ioannidis et al. 2006, 2005]. Groups of researchers (networks) exist that focus their research toward specific diseases. The network allows for individual researchers to pursue their own hypotheses, but it can also facilitate researchers by improving the quality of the studies through standardization of clinical, laboratory, and statistical methods. The Network of Investigator Networks helps to establish a roadmap for the conduct and translation of human genome research. The National Cancer Institute has funded different consortia of researchers to investigate specific cancers through their Epidemiology and Genetics Research Program.

3.5 Use of Banked or Stored Specimens

Human biological material is an essential tool for genetic research in humans. Demand for human specimens is increasing as more genetic research is conducted [Anderlik 2003]. One of the biggest challenges in conducting human-based research is recruiting participants and obtaining specimens. Large biobanks that contain stored specimens have been suggested as a solution to the problem [Nederhand et al. 2003]. Other sources of biological material could be specimens left over from other studies or even residual medical tissue that might remain after an operation, such as a biopsy. Technical issues present some challenges in using stored or banked samples. Specimens from epidemiological and biomarker studies are being collected by the thousands in ongoing studies. Little has been published on the selection and validation of the methods used to collect, prepare, preserve, and store these specimens [Holland et al. 2003]. Selection of these methods can affect the outcome of the study results and can even result in the specimens being useless for some analyses [Holland et al. 2003]. Development of a good quality control program using standard operating procedures when handling or storing the samples can help to ease concerns about sample integrity. Besides the technical issues surrounding stored or banked specimens, study design issues, such as selection bias, and ethical issues are also critical. These issues will be discussed in later chapters.

3.6 Cell Lines and Transgenic Animals

An alternative or complementary approach to epidemiology studies aimed at better understanding of disease pathologies in gene-environment interactions is the use of cell lines and transgenic animals, coupled with DNA microarray technology. This technology has expanded the field of genetic disease research to include evaluation of not just the gene, but also interindividual and intraindividual differences in gene expression. It is feasible to expose normal cells and tissues in vitro to chemicals or to complex mixtures of interest under controlled conditions and simultaneously to monitor exposure response by tens of thousands of genes. If this is done in the context of specific inherited genotypes of interest, then the underlying early exposure response can be identified.

The development of transgenic or genetically engineered animals is predicated on molecular epidemiological studies. These animals are widely used as basic research models to study the function of specific genes, the toxicity to specific chemicals, the result of a genetic change, etc. Candidate genetic traits or specific alleles can be introduced into an appropriate rodent model to predict or confirm other findings. In many cases, the animal will be susceptible to the disease/phenotype or partial phenotype only when the human homologue is present. These in vivo models, together with molecular epidemiological studies, are tools to understand disease pathobiology, develop prevention/intervention strategies, and provide data that will help the Occupational Safety and Health Administration (OSHA) develop better standards to protect workers.

3.7 Considerations in the Incorporation of Genetics Into Occupational Health Research

The collection of genotype data from workers may yield new insights into relationships between exposures, susceptibility, and disease, assuming that the analytical validity of the genotyping, selection bias, confounding, and interaction (i.e., gene-environment, gene-gene, and other) are adequately addressed. The rights of workers who volunteer to participate in research, including genetic research, must be protected. This protection must minimize the potential for misinterpretation, misuse, and abuse of genetic information by addressing issues such as privacy, confidentiality, notification, and the implications of the results for workers and their families. Providing the worker with a clear explanation of these and other aspects of a study during the informed consent process is the cornerstone of this protection.

The use of genetic biomarkers to improve the design and analysis of studies of occupational and environmental determinants of disease may be one way to address the limitations of observational epidemiology that have been described [Davey Smith 2001; Taubes 1995]. It is possible to exploit the random assignment of genes as a means of reducing confounding in exposure-disease associations through the application of Mendelian randomization principles [Davey Smith and Ebrahim 2003]. According to Mendel's second law, the random assortment of chromosomes at the time of gamete formation results in random associations between unlinked loci in a population and is independent of occupational and environmental factors. This, in theory, would lead to a similar distribution of unlinked genetic loci in individuals with and without disease [Little and Khoury 2003]. However, there are caveats to this approach. Attention must be paid to study size, differences in patterns of linkage disequilibrium, knowledge of candidate gene function, and the effect of population stratification [Little and Khoury 2003]. Early results from HapMap indicate a tendency of certain areas of the human genome to be inherited in large blocks rather than independent alleles [IHC 2005].

A rigorous study design can help to minimize biological variation. The contribution of the variation in biological changes in an individual or among individuals can be factored in if the characteristics and confounders of a genetic marker have been established. Protocols should be established for the collection and documentation of the specimens. The timing of the collection of the specimen may be critical, depending on what the test is measuring. Protocols also need to be developed to establish transportation and storage procedures in the preanalysis phase. It is generally accepted that specimens should be coded so that the identity, exposure status and disease status of the specimen donors are not known to the analyst. Questionnaires or interviews are needed to find out exposure details and about nonoccupational exposures that an individual might have, such as smoking, diet, or other lifestyle factors. It is critical in population validation that exposure assessment receive as much attention as marker measurement [Rothman 1995].

CHAPTER 4

HEALTH RECORDS: A SOURCE OF GENETIC INFORMATION

Genetic information may be included in the health records of workers [Anderlik and Rothstein 2001; Rothenberg et al. 1997]. This information can be in the form of a family history, inferences based on early age of onset of diseases in family members, a history of diseases with known strong genetic etiologies, or the results of physical examinations and common laboratory tests. This type of information is often routinely reported by workers on job applications, health questionnaires for jobs, insurance applications and physicals, and workers' compensation forms.

Genetic information is a particularly sensitive subset of health information, because it reveals distinctive and immutable attributes that are not just personal but shared by family members as well [Hustead and Goldman 2002]. Employers and prospective employers, who obtain family health history, gain some insight, albeit limited, about a worker's or potential worker's genetic make up, the genetic make up of his or her family members, and possibly his or her future health. However, the potential for misinterpretation and, hence, misuse of such information still exists.

The line between genetic and nongenetic information in health records is unclear. In existing and proposed legal statutes, the definition of genetic information may or may not include much of what is in health records other than that pertaining to explicit genetic tests [Hodge 1998]. Some statutes are underinclusive, limiting the definition of genetic information to DNA-based genetic test results. Other statutes are overinclusive and may include all types of genetic information as well as nongenetic health information [Hodge 1998]. A position statement of the American College of Medical Genetics (ACMG) noted that definitions must be sufficiently broad to accommodate the wide range of what is known about classic single-gene disorders and the contribution of multiple genes to common, complex diseases [Watson and Greene 2001; Williams 2001].

Workers may benefit from a better understanding of how genetic information can be used in the workplace. This information may be used to aid potential employees in employment decisions when science is able to provide answers with greater certainty about potential risks given the exposure in the prospective workplace. If workplace exposures are not an issue, genetic discrimination still may be. Employers who learn about rare monogenetic diseases (e.g., Huntington's disease) in a prospective worker's family may conclude that the worker has an increased risk and may thus hesitate to hire him or her. In 2000, a U.K. government-appointed committee ruled that British life insurers could use the results of genetic tests for Huntington's disease in underwriting life insurance policies [Aldred 2000]. However, due to fear of discrimination of those who test positive for Huntington's, other countries have introduced legislation that prevents insurers from either requiring genetic testing or of using the results of genetic testing to underwrite medical, disability, or life insurance policies.

4.1 Health Inquiries and Examinations

The Americans With Disabilities Act of 1990 (ADA) defined the kind of health information, including genetic, that an employer might obtain at three stages of the employment process:

- Preemployment, preoffer
- Postoffer, preplacement
- Postemployment, postplacement

In the preemployment, preoffer stage, the ADA prohibits health inquiries, including genetic inquiries and physical examinations, prior to extending a conditional job offer to an applicant [Langer 1996]. At the postoffer, preplacement stage, an employer could conduct unrestricted health inquiries or physical examinations, including genetic testing and inquiry. However, under the proposed regulations for the Genetic Information Nondiscrimination Act of 2009 (GINA) [FRN 2009], employers will be prohibited from obtaining family medical history or genetic tests of job applicants after making a job offer. Similarly, employers will be barred from offering genetic information through fitness for duty examinations.

4.2 Confidentiality, Privacy, and Security

Health records of workers and job applicants have been collected and maintained over time with different degrees of confidentiality. Confidentiality describes the duties that accompany the disclosure of nonpublic information, such as the release of health history records to a third party within a professional, fiduciary, or contractual relationship [Anderlik and Rothstein 2001]. In occupational health, traditional rules of confidentiality are often complicated by the dual roles of a health care provider who is involved in relationships with both the worker and the employer [Tilton 1996].

Underlying the responsibility of confidentiality is the right to privacy. The concept of privacy is broad and subsumes at least four categories: access to persons and personal spaces, access to information by third parties, third-party interference with personal choices, and ownership of materials and information derived from persons [Anderlik and Rothstein 2001; Allen 1997]. Privacy is linked to autonomy or individual self-governance, and it is an important theme in U.S. law and ethics [Beauchamp and Childress 1994].

A third concept, security, is related to confidentiality and privacy. It refers to the measures taken to prevent unauthorized access to persons, places, or information [Anderlik and Rothstein 2001]. The measures used to achieve the goal of security depend on the context and the state of the technology.

With regard to the health records of workers or job applicants, the issues of privacy, confidentiality, and security depend on a variety of factors. These include how the information is obtained, by whom, and for what purpose. Information can be records or test results, either of which may have genetic implications. A policy statement by the American College of Occupational and Environmental Medicine (ACOEM) asserts that the ethical standards of occupational medicine practice mandate that employers be entitled to counsel about an individual's medical work fitness, but not to diagnosis or specific details [ACOEM 1995]. Similarly, a position statement by the American Association of Occupational Health Nurses, Inc. (AAOHN), states that workers "should be protected from unauthorized and inappropriate disclosure of personal information" [AAOHN 2004b]. Protection of individual privacy and confidentiality of health information are also addressed by both organizations in their codes of ethics [ACOEM 2005; AAOHN 2004a].

Laws related to confidentiality of health information in the workplace are varied [Brandt-Rauf and Brandt-Rauf 1997]. ACOEM has endorsed a statement by the National Conference of Commissioners on Uniform State Laws that recommends "the development of uniform comprehensive legislation addressing the confidentiality of medical records...that encompass the treatment of employee medical information in the workplace" [Brandt-Rauf and Brandt-Rauf 1997]. Prior to passage of the ADA, employers not covered by the Rehabilitation Act of 1973 could use health information, including genetic information, as justification for not offering an applicant a job. Up to that time, health information was by norm or law generally kept confidential. However, depending on the relationship between the person who obtained the health history and the employer, a person's health records could be maintained in more than one location, including with the company physician, the human resources department, or a contract health care provider. Others, such as health and life insurance providers, could also have access to a worker's or job applicant's health history. Prior to 1990, use of genetic information was permissible unless otherwise prohibited by state law, either a disability nondiscrimination law or a genetic nondiscrimination law.

The handling of health records in occupational health may be influenced by various federal and state regulations that govern the release of private health information in the workplace. Some of these statutes include the Uniform Health Care Information Act [1986], workers' compensation statutes and case law [Workers' Compensation

Statutes 2007], the ADA [1990], the Drug Abuse Prevention, Treatment, and Rehabilitation Act [2007], and the Health Insurance Portability and Accountability Act (HIPAA) (Public Law 104–191) [HIPAA 1996]. HIPAA permits a group health plan or health insurer to request genetic information about an individual for treatment, payment, and health care operations, in accordance with HIPAA-compliant authorizations.

In 2000, the U. S. Department of Health and Human Services (DHHS) issued the Standards for Privacy of Individually Identifiable Health Information or Privacy Rule (45 CFR 160, 164) [DHHS 2007], which was amended in 2003 for compliance by April 14, 2003 [HIPAA 2005]. The rule imposed significant new documentation requirements on health care providers, particularly those who conduct clinical research or provide health data to researchers. Providers were required to obtain written consent covering the use or disclosure of personal health information.

The Privacy Rule did not replace or act in lieu of the DHHS protection of human participants regulations [DHHS 2007] and the Food and Drug Administration (FDA) protection of human participants regulations (21 CFR 50, 56) [FDA 1999]. The Privacy Rule does not apply to research; it applies to covered entities, which researchers may or may not be. The Privacy Rule may affect researchers because it may limit their access to information, but it does not regulate them or their research per se [NIH 2004]. This rule sets minimum standards for how protected health information may be used and disclosed and what control people have over their health information. For purposes of the Privacy Rule, genetic information is considered to be health information [NIH 2004].

4.3 Genetic Exceptionalism

"Genetic exceptionalism" is a term that describes the differentiation of genetic information from other health information and the contention that it should be afforded special protection [Kulynych and Korn 2002; Murray 1997]. However, definitions in state statutes vary widely and demonstrate the practical difficulty of drawing operationally meaningful distinctions between genetic and other types of health information. Broader definitions may expand genetic information in the future [Burris et al. 2000]. A compilation of state statutes on genetic privacy and genetic information can be found at the National Conference of State Legislatures [2004] Web site (http://www.ncsl.org/public/leglinks.cfm).

The concept of genetic exceptionalism is based on the implications that genetic information has for possible future illness of individuals as well as their family members. However, treating genetic information as distinct from health information may not be supportable because (1) genetic information may significantly overlap with other health information, (2) the issues underlying privacy and antidiscrimination protections for genetic information may apply equally to other sensitive types of health information (such as whether an individual is infected with the human immunodeficiency virus [HIV]), which may go relatively unprotected by comparison, and (3) the creation of significant protections for genetic information may foster a public perception that genetic information is something to hide rather than a valuable part of an individual's clinical record [Hodge 1998]. Ultimately, most diseases have a genetic component.

The literature contains unresolved arguments for and against claims of genetic exceptionalism [Launis 2000; Strudler 1994]. Some have argued that whether genetic exceptionalism is a valid construct is simply not the right question [Press and Burke 2001]. They argue that issues pertaining to genetic information should be viewed as central to and inextricably entwined with how disease, health prevention, and workplace responsibility are considered. Genetic information is risk factor information that, whether it needs special protection or not, is part of the constellation of factors that can be used to understand occupational risks. However, legislators and other policymakers have embraced the concept of genetic exceptionalism for pragmatic and political reasons [Rothstein 2005].

4.4 Genetic Discrimination

Family history has long been viewed as an important factor in health diagnosis and treatment [Andrews 1997]. In 2000, the EEOC published "Policy Guidance on Executive Order 13145: To Prohibit Discrimination in Federal Employment Based on Genetic Information" [EEOC 2000]. The Executive Order directs departments and agencies to extend to all of their workers the policy against genetic discrimination based on protected genetic information. The EEOC guidance directs that "protected genetic information" includes information about genetic tests on individuals or their family members and family health history [EEOC 2000]. A broader discussion of genetic discrimination appears in Chapter 7.

4.5 The Historical Use of Genetic Information

Prior to the availability of genetic testing, genetic information was available and was used by clinicians, employers, health and life insurers, and researchers. Phenotypic traits, such as race, ethnicity, and sex, have been widely used as crude factors to account for the potential influence of genetic factors (among others) in clinical and epidemiological studies. Seventy years ago, investigators speculated that genetic traits might predispose some workers to occupational disease [Haldane 1938].

Concerns pertain to the repercussions of genetic information falling into the wrong hands and resulting in loss of health and life insurance, loss of employment, having a mortgage foreclosed or denied, or having genetic information used in divorce and child custody cases, personal injury, or workers' compensation suits. No solid source of empirical evidence documents how often or for what purpose employers have historically obtained or currently obtain genetic information about job applicants. A 1999 American Management Association (AMA) survey reported that 24% of major U.S. firms collected information about family health history [AMA 1999]. By 2004, this had decreased to 14.7% [AMA 2004]. Furthermore, 4.2% of the surveyed companies in 2004 used family medical history in their decisions to hire, assign or reassign, or retain or dismiss employees. Attention has also focused on the privacy rights of family members and on who owns family health history data obtained from study participants [Renegar et al. 2001]. The issues involved are complex.

The American Society of Human Genetics (ASHG) [2000] has addressed the issue of family consent in research. ASHG believes that the determination about whether to collect family health history information represents more than minimal risk and affects the participants' rights and welfare and will have to be made in each case, as research protocols are reviewed at the local level. This determination will most likely be done by institutional review boards (IRBs). An IRB is a group that has been formally designated to review and monitor biomedical and behavioral research involving human participants. An IRB performs critical oversight functions to ensure research conducted on human participants is scientific, ethical, and legal. Most Federal government, academic and institutional (hospitals) groups have IRBs that review research studies conducted by their staff. Commercial IRBs are also available for industry supported research or other groups that may not have their own IRB [Lemmens 2000]. Thus, IRBs are well positioned to address the question of whether collecting family history data indeed represents a violation of privacy of living relative about whom information is collected.

4.6 The Use of Genetic Information in Research

Throughout U.S. history, health and occupational records and archival tissue of generations of workers have been an irreplaceable source of new knowledge about occupational diseases and their prevention, control, and treatment. Medical knowledge and occupational health cannot advance without ready, albeit controlled, access to health and occupational information. Health records may be overlooked and overshadowed as sources of genetic information by the potential power of DNA testing to yield genetic information. However, increasingly, health records may contain information about tests conducted for diagnostic, therapeutic, predictive, or other purposes. As health knowledge about diseases and genetic factors increases, genetic tests that were viewed as predictive may come to be considered diagnostic. As of 2008, use of genetic information contained in health and other records has not been studied and reported in the scientific literature, but it appears, at this time, not to be a major source of health information for occupational research.

4.7 Genetic Information in the Assessment of Causation

Genetic information has been used retrospectively to assess causation in workers' compensation or tort litigation [Schulte and Lomax 2003; Poulter 2001]. Many compensation or tort litigation cases that have invoked genetic susceptibility in relation to toxic or other causation have used the argument that genetics is an alternative explanation to the toxicant for the causation of disease and that a family history of the disease or condition in question is evidence of genetic predisposition [Poulter 2001]. Counterarguments often invoke the absence of a family history, suggesting the absence of genetic expression in the absence of occupational exposure.

In some instances, genetic information may be used to apportion causation. The extent to which genetic information from a health record can be used to apportion causation will depend on the type of information in the record. Family history, genetic test results, or medical test results with genetic implications may all contribute to a causal analysis, but not to the same extent. Information about mutations related to monogenic diseases that are highly penetrant may provide the strongest evidence for gene causation. However, most monogenic conditions are not work-related. In contrast, with less-penetrant polymorphisms, which may be involved in the activation or detoxification of workplace chemicals, the strength of arguments based on such information could be limited because of low relative and attributable risks related to the polymorphisms. [Kelada et al. 2003; Marchant 2003b; Marchant 2000].

CHAPTER 5

GENETIC MONITORING: OCCUPATIONAL RESEARCH AND PRACTICE

Genetic monitoring is the evaluation of an exposed population for genetic damage over time and involves the detection of biomarkers of effect (see Table 2–2). Genetic monitoring has been used in a variety of situations, particularly involving radiological and genotoxic chemical exposures. It is similar to other forms of biological monitoring [Ashford et al. 1990; Schulte and Halperin 1987]. Little information is available about the recent use of genetic monitoring by companies. The last available information was a survey of Fortune 500 companies conducted by the now defunct Office of Technology Assessment (OTA) in 1989. Only one company reported current use of genetic monitoring [OTA 1990]. Five companies reported past use, and two companies reported consideration of future use of genetic monitoring.

In public health, genetic monitoring has many of the same strengths and limitations as toxic exposure or health effects monitoring. The strengths of a monitoring program include identifying a risk of exposure for a group or possibly for individuals to potentially hazardous substances, targeting work areas for evaluation of safety and health practices, and detecting previously unknown hazards. When genetic monitoring is used to evaluate an exposure or the effectiveness of safety and health practices, a reduction in the genotoxic measurements indicates a reduction in exposure. If the genetic monitoring assay is being used to determine a cancer risk, a reduction in genotoxic measurements may indicate a reduction in cancer risk, but only if that marker is in a relevant pathway for cancer. Genetic monitoring has its limitations, e.g., it will be uninformative if the exposures do not cause genetic damage. A number of confounders can also affect the results and produce large variations in measurements.

Genetic monitoring has been used to quantify radiation exposures in military and civilian workers handling nuclear materials in the United States and elsewhere in a research setting [Blakely et al. 2001; Jones et al. 2001; Moore and Tucker 1999; Mendelsohn 1995; Langlois et al. 1987]. A case study of monitoring Chernobyl workers is presented.

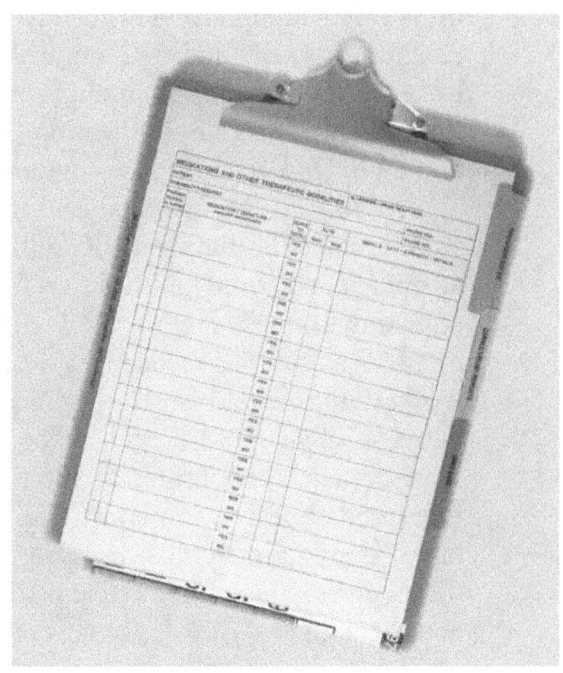

Case Study: Genetic Testing in Chernobyl Cleanup Workers

An estimated 600,000 to 800,000 workers in the former Soviet Union were potentially exposed to ionizing radiation as a result of cleanup activities after the Chernobyl nuclear power reactor accident in 1986. Approximately 119,000 cleanup workers were monitored for radiation exposure using traditional badge monitoring methods, and those who were monitored in the 2 years after the accident received an annual rate of exposure of about 0.75 to 1.5 times the current annual U.S. exposure limit for workers exposed to gamma radiation.

A collaborative research program was established between U.S. and Russian scientists to evaluate the potential utility of persistent genetic biomarkers of exposure and damage among these badged workers. The goal of this research was to develop assays that could (1) estimate exposure, (2) indicate whether genetic damage had occurred that might lead to harmful health effects, and (3) estimate the relative biological effect of different types of radiation exposure. One potential public health use of such research is the ability to establish better regulatory standards by permitting the comparison of doses among populations from which standards may be derived (e.g., Japanese atomic bomb survivors versus nuclear workers) and by establishing the relative harm caused by different forms of radiation and different rates of exposure.

The studies conducted in the Chernobyl population have shown variability in their ability to assess genetic damage and to confirm the doses measured by badge readings (reviewed in Jones et al. [2002]). A study among 625 cleanup workers and 182 controls from Russia compared three assays of genetic damage, including chromosomal aberrations measured by fluorescence in situ hybridization (FISH), HPRT mutation in lymphocytes, and two GPA-variant flow cytometry assays using erythrocytes [Jones et al. 2002]. Extensive analysis of other exposure and lifestyle attributes were included in the study. Results indicated that age, smoking, and other attributes and behaviors were related to markers of genetic damage. After adjusting for these factors, the FISH and HPRT assays indicated an average of 30% and 41%, respectively, greater genetic damage among cleanup workers than controls. The GPA assays showed no significant difference between cleanup workers and controls. The doses measured by badges were in general confirmed by the FISH chromosomal assay, and the effects measured appeared to persist over time. A number of technical problems have been raised by this research, including the assay sensitivity at lower doses, interactions with age and lifestyle factors, and technical difficulties and the high cost of these assays for use in epidemiological research.

In the early 1960s, the Texas Division of Dow Chemical Co. initiated a comprehensive surveillance program for workers exposed to a variety of chemicals [Kilian and Picciano 1979]. The purpose of the program was to evaluate whether cytogenetic changes could be used to monitor for exposure to chemicals. The program consisted of four parts: (1) periodic evaluation (health history, physical examination, and laboratory reports), (2) cytogenetic evaluation by assessing chromosomal aberrations, (3) conventional epidemiology such as morbidity and mortality, and (4) nested case-control studies to look at specific diseases or conditions. Although this program was eventually phased out by the company, some results were achieved, such as the development of a comprehensive health questionnaire to determine confounders and interferences with cytogenetic results; the development of an education program to explain the program to the workers; the accumulation of a large database of cytogenetic results that could show temporal and seasonal variations of cytogenetic tests; and the confirmation that benzene and epichlorohydrin were genotoxic to workers who were exposed to either chemical [Legator 1995].

Biomarkers of genetic exposure or damage can be sources of risk estimates to be used along with morbidity and mortality statistics. The use of well-validated genetic biomarkers has been advocated as a way to prioritize exposed individuals for more thorough medical monitoring [Albertini 2001]. However, one concern raised by critics of genetic monitoring is that the resultant action will focus only on the worker and not on minimizing or eliminating the exposure [Ashford et al. 1990; OTA 1990; Schulte and Halperin 1987]. Depending on the circumstances, both medical monitoring of the worker and decreasing the exposure may be warranted. The hierarchy of controls (Table 5–1) and the principles of the OSH Act require emphasis on changing the workplace environment to control occupational exposures at the source. Primary prevention tools are substitution or elimination of the agent, engineering controls, training, administrative controls, and personal protective equipment [Ellenbecker 1996; Halperin and Frazier 1985].

Secondary prevention includes medical monitoring and controlling other exposures that are likely to contribute to a health risk. Medical monitoring is one approach for detecting problems early enough to make a difference in the natural history or prognosis of the health risk and may be justified if unidentified exposures exist or if exposures cannot be adequately controlled. Genetic monitoring can be a tool used in primary or secondary prevention strategies to assess the efficacy of exposure or effect reduction.

In the course of genetic monitoring, individual results have prompted a decision for further prevention measures such as medical removal—moving the worker to another job location where exposure is lower or absent. This has been practiced in conjunction with traditional biological monitoring, such as for blood lead and serum and urinary cadmium [OSHA 2007a, 2007b; 1974].

Genetic monitoring highlights confusion in the literature between group and individual risk assessment. Epidemiological research,

Table 5–1. Hierarchy of controls for occupational health programs

1. Primary prevention: Eliminate or reduce exposure
   - Elimination of substitution of the hazard
   - Engineering controls
   - Administrative controls
   - Work practices
   - Personal protective equipment
2. Secondary prevention: Reduce the biological effects of exposure
   - Medical monitoring
     Pre-exposure screening
     Detection of effects following exposure
        test body burden
        test genotoxic effect
   - Control of other chemical/agent exposures *that may contribute to the genotoxic effect*
3. Tertiary prevention: Reduce disease impact
   - Medical removal
   - Job reentry

*Italicized text added to show how genetic monitoring could fit into the hierarchy.
(Adapted from Halperin and Frazier [1985]; Plog and Quinlan [2002].)

including the validation of biomarkers of exposure and effect in a population, identifies risks only for groups. It does not identify risks for specific individuals. Individual risk profiles could be constructed using exposure factors, data from tests on effects of exposure (such as genetic monitoring), and hereditary characteristics [Truett et al. 1967]. However, epidemiological data reports are still primarily based on group findings. Thus, what is available is the risk for a specific group with certain characteristics in common, similar to life insurance company ratings of groups such as smokers. In practice, the individual risk profiles could be useful for risk monitoring to identify highly exposed workers. A concern about this kind of risk profiling is the possibility of removing or excluding workers who fit the profile, rather than correcting environmental exposures.

5.1 Regulation

Currently, no U.S. regulations mandate genetic monitoring, although medical surveillance requirements are included in 17 OSHA standards, but only 4 of these (arsenic, lead, cadmium, and benzene in emergency exposure situations) require specific biological monitoring to determine exposure. GINA does permit employers to engage in genetic monitoring provided it meets certain requirements such as providing clear written notice of the testing, that it be voluntary, why the testing is being done, employee consent is received and that the employee receives the results in such a way that the identity of specific individuals cannot be determined [FRN 2009].

Questions arise about whether genetic monitoring indicates a potential health problem, an existing health problem, or compensable damage [Schulte and DeBord 2000]. Further work to understand and interpret the science and public policy will help to answer these questions. No genetic monitoring assay has yet been fully validated to assess an individual's risk of disease. Changes in chromosomal aberrations have been associated with increased risk of developing cancer. However, no individual risk for cancer development has been identified. As additional research is conducted and as understanding and recognition of genetic changes relevant to exposure and disease increase, genetic monitoring assays may be able to provide information about individual risks. With this increased knowledge, new opportunities for detection, prevention, and treatment of occupational disease may arise.

5.2 Considerations for Genetic Monitoring

As with any medical monitoring program, a plan to delineate the objectives, scope, and resulting action of the program is needed. If genetic biomarkers are to be useful in the workplace, certain criteria need to be established. Examples of criteria proposed by Lappe [1983] and Murray [1983] are given in Table 5–2. Decisions need to be included in the plan, before implementation of monitoring, as to what will happen to workers with outlier results. The followup could range from repeat monitoring to diagnostic evaluation and may include environmental remediation or medical removal.

Since many genetic biomarkers can be influenced by nonworkplace exposures, lifestyle, and genetic makeup, genetic monitoring would be accompanied by a questionnaire to assess these exposures (which could include neighborhood ambient air, diet, hobby exposures, etc.). The goal of genetic monitoring would be to explain the implications of the monitoring results rather than to attribute them to some nonoccupational exposure.

Consideration of what workers would be told about the monitoring plan, their test results, and what those test results mean is important for the well-being of the worker. The question of who will have access to genetic monitoring data needs to be answered prior to implementation of any plan. Potential participants would be informed of how their data would be protected from access by others and how their data would be used.

Genetic markers of exposure or effect of exposure for occupational toxicants are not sufficiently validated at this time to permit their use in routine practice or regulation, except for genetic damage related to radiation exposure. However, a growing body of data links genetic biomarkers and somatic mutations in reporter genes to cancer risks in groups of workers [Boffetta et al. 2007; Bonassi et al. 2007, 2004, 2000; Rossner et al. 2005; Liou et al. 1999; Hagmar et al. 1998]. The relevance of this risk to individuals in those groups with increased levels of genetic biomarkers has not been established and needs further study. At this time, insufficient evidence exists to support genetic biomarkers in routine occupational safety and health practice or regulation.

However, the use of genetic biomarkers in research settings and in validation studies should continue to fill in those knowledge gaps and establish links between genetic biomarkers and occupational diseases, if they exist.

Table 5–2. Potential Criteria for Implementing Genetic Monitoring in the Workplace

- Use of validated genetic markers
- Clinical utility established
- Goals of the program specified
- Acceptance by population being monitored (informed consent)
- Established linkage to exposure or disease
- Protection of privacy and confidentiality
- Notification of participants
- Process for addressing results and outliers

(Adapted from Lappe [1983]; Murray [1983].)

CHAPTER 6

THE THEORETICAL USE OF GENETIC SCREENING AND OCCUPATIONAL HEALTH PRACTICE

In the workplace, genetic screening is the examination of the genetic makeup of workers or job applicants for certain inherited characteristics. This discussion of genetic screening is intended only for informational purposes and for stimulation of discussion. Genetic screening is not being recommended to inform employers in making employment-related decisions, and currently no test has been validated for genetic screening purposes in an occupational setting.

Theoretically, genetic screening in the workplace could be applied for two distinct purposes (Figure 6–1). First, workers or job applicants could be screened for the presence of genetically determined traits that would render them susceptible to a pathological effect if exposed to specific agents in the workplace. This information could also be used by the individual to make job-related choices. Second, workers or job applicants could be screened to detect heritable conditions associated with

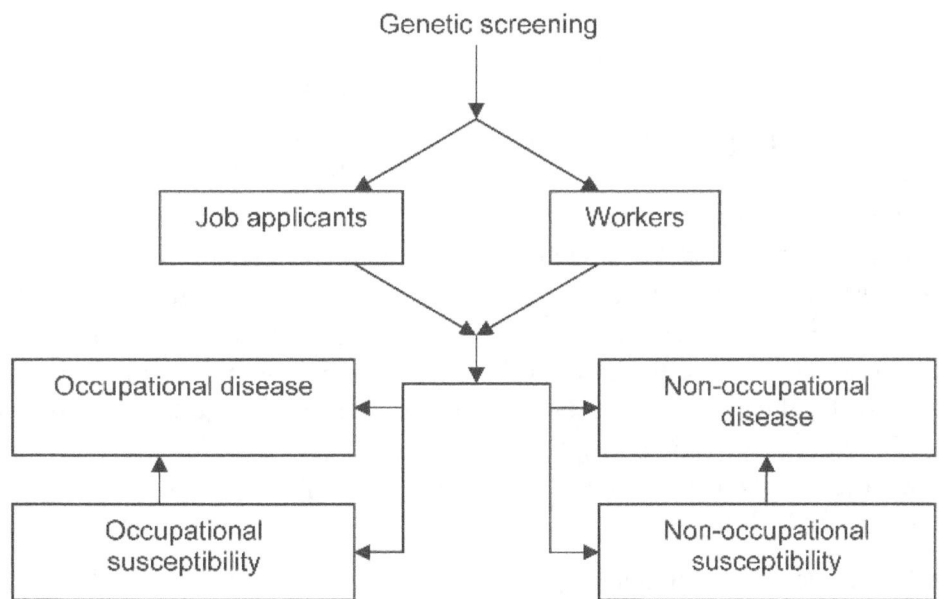

Figure 6–1. Theoretical uses of genetic screening. (Adapted from Bingham [1998].)

diseases unrelated to occupational exposure [OTA 1990].

6.1 History

Haldane [1938] is generally credited as the first scientist to suggest using genetic screening to identify and exclude susceptible workers from hazardous work environments:

> *The majority of potters do not die of bronchitis. It is quite possible that if we really understood the causation of this disease, we should find that only a fraction of potters are of a constitution, which renders them liable to it. If so, we could eliminate potters' bronchitis by rejecting entrants into the pottery industry who are congenitally disposed to it.*

Later observations of the genetic makeup affecting an individual's reactions to either a chemical agent or a drug were reported in the 1950s during the Korean War, when some American soldiers taking the antimalarial drug primaquine experienced acute hemolytic anemia. This was attributed to their carrier status for the gene for glucose-6-phosphate dehydrogenase (G6PD) deficiency. These soldiers were described as hypersusceptible [OTA 1990].

Schill [2000] describes the early history of genetic screening in the workplace:

> *G6PD is an enzyme necessary in the glucose metabolism of the red blood cell. Based on the experience of the soldiers in the Korean Conflict, it was postulated that individuals with a deficiency of G6PD also would develop acute hemolytic anemia after exposure to chemicals such as aromatic nitro and amino compounds; arsine and related metal hydrides; lead and its compounds; and several dye intermediates [Rothstein 1984]. Stokinger and Mountain [1963] published one of the first scientific journal articles advocating the use of genetic screening to identify individuals who were hypersusceptible to certain chemicals because of their G6PD deficiency. By the early 1970s, genetic screening for hypersusceptibility had been proposed for additional conditions, such as sickle cell trait, alpha-1-antitrypsin (AAT) deficiency, and carbon disulfide sensitivity [Rothstein 1984].*

The issue of genetic screening was first addressed by OSHA in the 1970s. At that time, OSHA promulgated 14 carcinogen standards that required a preassignment examination by a physician before a worker could be assigned to a job covered by these standards. The OSHA standards specify that the examination include a personal history of the worker, a family history, and the occupational background of the worker, including genetic and environmental factors [OSHA 1974]. No further guidance was developed, and initially the issue did not receive much attention [Schill 2000].

Subsequently, in February 1980, after the publication of a series of articles in the *New York Times* on genetic testing in the workplace, the issue of using genetic factors in the carcinogen standards received greater scrutiny [Schill 2000]. In response, Dr. Eula Bingham, then Assistant Secretary of Labor for OSHA, issued a news release that stated, "There is absolutely no OSHA standard that requires genetic testing of any employee" [OSHA 1980b]. Bingham said:

> *Exclusion of workers as a result of genetic testing runs contrary to the spirit and intent of the Occupational Safety and Health Act of 1970. It wrongly puts the burden of controlling toxic substances on the worker who is denied employment because of a supposed sensitivity. Employers should make the workplace safe for all workers, rather than deprive some workers of their livelihood in the name of safety.*

This announcement was followed by a directive from the OSHA Office of Compliance Programs to OSHA enforcement staff in 1980. The directive stated in part, "These provisions [of the carcinogen standards and the OSHA cancer policy] do not require genetic testing of any employee [or] the exclusion of otherwise qualified employees from jobs on the basis of genetic testing" [OSHA 1980a]. Furthermore, the directive explained that taking a worker's health history must be considered a "routine part of standard medical practice...designed to identify factors important to the employee's general health status" [OTA 1990; OSHA 1980a].

6.2 Past and Current Use of Genetic Screening

In the early 1980s, at the request of Congress, OTA studied the role of genetic testing in the prevention of occupational disease [OTA 1983]. As part of its evaluation, OTA surveyed U.S. industry, including utilities and unions, to determine the extent and nature of genetic testing (monitoring and screening) that was occurring in the workplace. The survey found that genetic testing had been used by 17 organizations in the previous 12 years, but only 5 of the 17 were conducting such testing at the time of the survey. However, 59 organizations expressed interest in future use of genetic testing [OTA 1983].

A second study of genetic testing in the workplace was undertaken by OTA in the late 1980s. This study found that 12 companies were using genetic monitoring or screening for research or some other reason [OTA 1990]. Six companies reported that they anticipated future use of genetic monitoring or screening [OTA 1990].

A 1999 American Management Association (AMA) survey on medical testing in the workplace found that 74% of 1,054 responding companies required physical examinations of newly hired and current workers [AMA 1999]. This survey determined the frequency and types of genetic testing of job applicants and workers, as well as how the test results were used

[Schill 2000]. Companies that tested used the results to hire applicants (6.7%), to assign or reassign workers (7.3%), to retain or dismiss workers (1.7%), and in any regard (10.3%) [AMA 1999]. The AMA repeated this survey in 2004 and found that 63% of the 503 companies surveyed required some type of medical testing [AMA 2004]. Testing categories included Huntington's disease, family medical history, and susceptibility to workplace hazards. The companies in the second survey indicated that the results from the aforementioned categories were used to hire job applicants (12.9%), assign or reassign employees (8.4%), or retain or dismiss employees (4.8%). The percentages from the second survey reflect slight increases from those of the first survey. However, since the companies were not identified in either survey, it is possible that different companies answered the two surveys so that the increases only reflect a difference in the respondents and not that more companies were beginning to use medical testing to make employment decisions.

An example to illustrate the potential use of genetic screening may be seen in the case of beryllium workers. Research has demonstrated that workers with beryllium sensitivity and CBD were more likely to carry $HLA\text{-}DPB1^{E69}$ than workers without these conditions, although the PPV was not very high. In the last few years, these findings prompted a beryllium manufacturing company to initiate a pilot preemployment screening program for prospective workers [Bates 2001]. Prospective workers were informed that a test for a genetic marker that had been linked to CBD ($HLA\text{-}DPB1^{E69}$) was available at no cost to them. Genetic counseling was provided by a university-based genetic counselor through an 800-number. The tests were performed at a university-based laboratory. The beryllium manufacturing company paid for the testing and counseling, but did not receive identifiable individual results. Because of economic conditions in the beryllium industry at the time of the pilot project, the company hired few new workers so that relatively few individuals used the program, and therefore it was suspended. In addition, evidence that an enhanced preventive model of workplace controls was effective in reducing exposure and rates of sensitization to beryllium resulted in less interest in the issue of individual susceptibility conferred by genetic status.

Although currently no genetic screening tests have been validated for assessing the increased risk of susceptibility to workplace hazards, it is anticipated that such tests will eventually become available for voluntary use. Genetic screening can offer some benefits, such as giving a person additional information about whether a job may affect his/her health, as in the beryllium case described above. At such time in the future, a test may be developed that would warrant its use to protect public safety. It has also been suggested that as genetic tests become more available in society, employers who fail to use them may be held liable for damages sustained by a worker whose genetic condition causes a lapse of consciousness or incapacity [French 2002]. This view has not been substantiated in the literature, nor is it plausible that a company would be held liable for genetic conditions of its employees. The implications are also not known if an employee knew about a genetic variation that could increase or decrease risk from exposure and failed to disclose that information to his employer.

6.3 Technical and Public Health Issues in Worker Screening

6.3.1 Reversing the hierarchy of prevention

Some proponents of genetic screening argue that, in a competitive business environment, employers seek to use technical innovations, such as genetic screening, to select workers. Genetic factors already affect employment options to the extent that they affect abilities. Ostensibly, organizations would use genetic screening to avoid placing hypersusceptible workers in hazardous jobs. With these potential benefits, some have argued that companies have an obligation to screen [OTA 1990].

As described in Chapter 5, the occupational safety and health community has established a hierarchy of controls for preventing occupational disease and injuries [Halperin and Frazier 1985]. Genetic screening is not part of the hierarchy of controls. It involves evaluating workers prior to employment. An evaluation of health history is part of job placement decision-making. The addition of genetic screening to the process of job placement has the potential to reverse the emphasis in the hierarchy of controls from changing the environment to changing (excluding) the workers [Schulte and Halperin 1987].

6.3.2 Uncertainty of the science and premature application of genetic screening tests

Various authoritative and scholarly groups have identified criteria for the use of genetic tests [Genetics and Public Policy Center 2006; ACOEM 2005; ASCO 2003; CDC 2003; Goel 2001; NBAC 1999]. In addition to the ADA requirements that tests be job-related and consistent with business practice, as well as other civil rights stipulations in legislation, there are the issues of the validity and utility of the tests, as well as ethical, legal, and social safeguards. In short, an adequate evidence base is needed prior to the use of a genetic screening test. Until such a base is established, the test would not be ready for use in a worker population. In 2005, ACOEM concluded in its position statement on genetic screening in the workplace that "until extensively validated, genetic screening is a form of human investigation and subject to the appropriate ethical and scientific controls."

6.3.3 Published criteria for genetic screening

In anticipation of the eventual use of genetic screening tests to determine the increased risk of susceptibility to workplace hazards, the scientific community has been engaged in the discussion and consideration of potential criteria that would support the use of genetic screening for employers or workers to make employment-related decisions. For example, the American Medical Association Council on Ethical and Judicial Affairs [1991] has suggested that genetic screening only be used if specific conditions could be met:

- The disease develops so rapidly that serious and irreversible illness would occur before monitoring of either the workers' exposure to the toxic substance or the workers' health status could be effective in preventing the harm.

- The genetic screening test is highly accurate, with sufficient sensitivity and specificity to minimize the risk of false negative and false positive results.

- Empirical data demonstrate that the genetic variation results in an unusually elevated susceptibility to occupational illness.

- Undue expense is needed to protect susceptible workers by lowering the level of the toxic substance in the workplace.

- Testing is not performed without the informed consent of the worker or applicant for employment.

Yesley [1999] also offered criteria to consider regarding exclusionary employment policies based on genetic screening for occupational susceptibility:

- The relative and absolute risk of the disorder if an individual with the susceptibility mutation receives the occupational exposure

- The accuracy of the genetic test in detecting the mutation

- The seriousness of the disorder

- The availability of treatments or preventives

- The practicability of eliminating the exposure from the workplace

Applying these criteria, Yesley [1999] provided three instructive examples: one that would reject the use of an exclusionary policy based on genetic screening, one that would support the use of an exclusionary policy, and one that would fall in the gray zone (Table 6–1). Other criteria may also include the public safety or worker safety aspects of a job. At present, however, exclusionary employment-related policies based on genetic screening for increased risk to workplace hazards are not justified because science has not shown definitive linkage of genes and occupational illness. In addition, no genetic screening test with regard to an occupational illness has been validated.

Table 6–1. Exclusionary policies and genetic screening [Yesley 1999]

Exclusionary Policy	Criteria
Not justified	• It is practical to eliminate the hazardous exposure.
	• A polymorphism does not substantially increase the risk from an occupational exposure.
	• The disorder is mild, slow to manifest, and treatable.
Gray zone	• A polymorphism confers susceptibility to an occupational exposure, resulting in a serious but nonfatal disorder that will not manifest for several years and may be treatable.
	• The exposure can be eliminated at reasonable cost.
May be justified	• A polymorphism in combination with an occupational exposure commonly causes a fatal disorder that never occurs in the absence of the mutation.
	• Control of the exposure from the workplace is not practicable.

CHAPTER 7

THE ETHICAL, SOCIAL, AND LEGAL IMPLICATIONS OF GENETICS IN THE WORKPLACE

Although genetic research is generally no different from many other types of biomedical research, the information obtained in such research may have a greater potential for misinterpretation, misuse, and abuse. These same concerns also pertain to the use of genetic information in occupational health research and practice. This chapter highlights some of the ethical, legal, and social issues related to genetics in the workplace.

It has long been known that there is a range of variability in human response to occupational hazards, particularly chemical hazards. Genetic factors contribute to the variability and consequently may be useful to consider in research and control of hazards [Marchant 2003b; Christiani et al. 2001; Neumann and Kimmel 1998]. "Susceptibility" is a term used to describe different ways that individuals respond to occupational and environmental contaminants. The concept of susceptibility in occupational safety and health should be framed within a public health context [Froines et al. 1988]. The basis for this framework is that workplace disease and injuries represent one of the largest groups of preventable conditions in public health, and the most effective strategies for the prevention of disease and injury are primary prevention strategies. Historically, excluding susceptible workers from exposure has not been considered an element of primary prevention. The social history of regulation and practice has been to emphasize control of the workplace and not the workers [Froines et al. 1988].

Identifying susceptible individuals or groups before exposure might enable those individuals to be protected. However, such policies might also result in discrimination and economic hardship independent of the disease that the policy was intended to prevent [Froines et al. 1988]. Furthermore, identification of susceptibility might be interpreted as a control strategy in itself, in conflict with historical values and approaches exemplified in the OSH Act. Employers may also face an ethical dilemma if they expose a known susceptible worker when the workplace exposure cannot be controlled enough to protect that worker [Froines et al. 1988].

The economic issues of employers and workers also influence the application of genetic susceptibility in occupational safety and health policies and practices. Employers face increasing health care costs, and often a small fraction of the worker population accounts for a large portion of the health care expenditures. There is a strong incentive for employers to try to reduce this fraction. From the workers' perspective, current or prospective workers' livelihoods depend on having equal opportunities to obtain and keep jobs and to flourish in their work. Workers also have the expectation to be protected from workplace harm and compensated for adverse health effects from work.

Implications for Occupational Safety and Health

In spite of the OSH Act and the historic view of workers' compensation as a no-fault system, companies may argue that their legal liability should be lessened because a worker's genetic makeup contributed to a disease. This argument may be given more credence if the worker elected to take a genetic test that showed increased susceptibility, but chose to accept the job anyway. However, the dynamics involved in job selection are complex, and the OSH Act does not imply that susceptible workers should be unprotected. How society will address the disease burdens and costs related to susceptibility will depend not only on economic analyses but also on prevailing political views of distributive and social justice.

7.1 Framework for Considering Genetic Information

Genetics in the workplace can be considered according to three categories of use: research, practice, and regulation/litigation. All of the uses of genetic information in the workplace can be viewed through these three categories. To further explore these categories, they will be considered in terms of inherited genetic factors and acquired genetic effects. This is a common classification scheme for genetic risks. Inherited genetic factors pertain to germ and somatic cell DNA transmitted through meiosis or mitosis. Acquired genetic effects involve modification of genetic material over time and can include genetic damage or expression as a result of workplace and environmental exposures.

The line between inherited genetic factors and acquired genetic effects, however, can be blurry in some areas, particularly those related to gene expression status such as transcriptomics, proteomics, toxicogenomics, and metabonomics. Table 7–1 identifies some of the ethical, legal, and social issues for each use of genetic information.

7.2 Inherited Genetic Factors: Research

Genetic factors are likely to be responsible for some differential distribution of diseases among workers that cannot be accounted for by differences in exposures and lifestyle. [Neuman and Kimmel 1998]. It is clearly accepted that practically no occupational diseases are determined solely by either genes or environment. In the early history of occupational epidemiology, genetic influences were considered only in terms of confounding by race and sex. Today as many occupational exposures are being controlled to lower levels, the understanding of genetic factors as sources of variability in risk estimates is increasing [Vineis et al. 1999; Neumann and Kimmel 1998].

7.2.1 Safeguarding rights of participants in research

Genetic research involves human participants, and the rights of these participants require protection. The cornerstone of protecting the rights of research participants is the informed consent process, which is based on three historic documents: the

Table 7–1. Framework for Considering Ethical, Legal, and Social Issues of Genetics in the Workplace With Respect to Sections in the Chapter

	Types of genetic information	
Uses	Inherited genetic factors	Acquired genetic effects
Research	Section 7.2 • Validity and predisposition • Safeguard rights of research participants • Interpretation and communication of results of occupational genetics research	Section 7.5 • Validation • Interpretation and communication • Justice • Privacy
Practice	Section 7.3 • Prevention and diagnosis • Job actions • Autonomy, privacy, and confidentiality • Stigmatization and discrimination • Validation	Section 7.6 • Use in genetic monitoring • Interpretation • Validation • Use in genetic screening
Regulation Ligitation	Section 7.4 • Premature use of genetic tests • Apportionment of causation • Hypersusceptibility in regulations	Section 7.7 • Prescreening chemicals • Use in risk assessment • Impact on risk management

Implications for Occupational Safety and Health

Nuremberg Code [1949], the Belmont Report [1979], and the Federal Policy for the Protection of Human Subjects, also known as the Common Rule [1999], which is codified at 45 CFR 46, Subpart A [DHHS 2005]. These documents form the basis for protecting the rights of participants in biomedical research.

A broad spectrum of opinion exists about what obtaining informed consent entails and when it is achieved [Clayton 2003; Schulte et al. 1999; Samuels 1998a; Hunter and Caporaso 1997; Schulte et al. 1997]. Some believe that for genetic data (biomarkers) whose meaning is not known at the time of the study, a participating worker in an occupational study cannot give truly informed consent [Samuels 1998a]. This interpretation implies a much higher standard for genetic biomarker information than for other information routinely obtained by questionnaires, environmental monitoring, or record linkage. Until there is determination of predictive value and course in the natural history, such genetic biomarkers are clearly only research variables with no clinical meaning, and participants should be made aware of this. The extent to which a biomarker has been validated (i.e., quantitatively linked to risk of disease at the group or individual level) should be clearly described to potential research participants. With regard to informing participants of risks, general practice has been to identify only medical risks; however, it has been argued that truly informed consent should include reference to non-medical risks that might affect participants. For example, study participants may be informed that they carry a genetic mutation that puts them at increased risk of subsequently developing cancer given a particular exposure. Participants in occupational genetic studies consent to provide the specimens and corollary demographic and risk factor information and, hence, cooperate in the specified research. The participant generally does not consent or imply consent to distribution of the data in a way that identifies him or her individually to any other parties, such as employers, unions, insurers, credit agencies, lawyers, family members, public health agencies, etc. [Schulte et al. 1997].

Many of the ethical concerns that have arisen with single-gene studies will be exacerbated as investigators conduct genome-wide association studies (GWAS) in large cohorts or in combinations of cohorts [IHC et al. 2007; Hinney et al. 2007; Nahed et al. 2007; Weir et al. 2007]. The utility of the GWAS approach will be maximized because the genetic data are posted in widely accessible databases [Couzin and Kaiser 2007]. While such approaches may be powerful research tools and resources, they have the potential to allow an individual in the database to be identified. Consequently, the privacy of individuals in such large databases is in jeopardy. Underlying the privacy issues is the nature of the original informed consent, the safeguards in the database assembly procedures, and the limitations on the use of the database by other investigators. In addition to privacy and consent issues is the potential for developing premature and unvalidated clinical guidance based on findings of GWAS or the use of such findings in litigation or criminal proceedings.

The broad range of opinion of professional organizations and scholarly groups is that genetic testing in research or practice should only be conducted when it is voluntary and given with informed consent [ACOEM 2005]. One question that arises with genetic testing for research in the workplace is whether a prospective or current worker can freely give consent. The power dynamics of a workplace are such that consent could easily be pressured or coerced even if coercion is not intentional [Samuels 1998b].

Genetic research has some special aspects that relate to the informed consent process. New technologies may come into existence after the specimens have been collected and informed consent has been obtained. Ethical issues for stored specimens relate to whether (1) consent was originally given to store the samples, (2) the consent was generic or specific to the original hypothesis, (3) the original consent obtained would meet consent standards at the time specimen use was contemplated, and (4) results might pertain to family members who were not part of the informed consent process [Schulte et al. 1997]. Obtaining reconsent is difficult for several reasons: study participants are hard to recontact, bias may be interjected into the study if a high proportion of the participants deny reconsent, and if reconsent is needed for every new test then a continual process of reconsent may be necessary as new assays are developed. Development of common informed consent language to allow testing of specimens using yet-to-be-developed tests is not likely to be successful. Informed consent usually limits testing to the specific hypothesis at hand. What has been successful is adding language to informed consents to allow storage of specimens to address future questions related to the specific disease at hand as opposed to storage of samples to address any hypothesis. The emergence of commercial tissue banks may help to alleviate the issue in that few restrictions may exist to limit hypothesis testing. However, broad social and policy questions surround these commercial banks with respect to informed consent, privacy, and the potential for genetic analysis to generate a large amount of information from a small specimen [Rothstein and Knoppers 2005; Anderlik 2003].

Some investigators have drafted language to ask for consent to anonymize any unused specimens. Anonymization may allow the investigator to use those specimens for other testing or methods development. Anonymization comes with a price, however, in that valuable information about the person who donated the specimen, such as exposure or work history, may be lost. In addition, anonymization does not allow the investigators to notify the participants if a clinically relevant finding is observed, since the identity link between the specimen and its donor has been destroyed.

7.2.2 Interpreting and communicating the results of occupational genetics research

Three issues merit consideration in the interpretation and communication of the results of genetic research. These are the realization that (1) epidemiologic results are group risks and not individual risks, (2) a statistically significant genetic factor may not be biologically significant, and (3) the results of many small studies of genetic polymorphisms have not been replicated [Schulte 2004]. Many gene disease association studies represent new findings and have not been replicated by other investigators; and therefore, do not offer a clear clinical interpretation [Renegar et al. 2006]. Heeding these issues, a CDC multidisciplinary group [Beskow et al. 2001] using expert opinion, as well as federal regulation, the National Bioethics Advisory Commission's (NBAC) report on research involving human biological materials [NBAC 1999], and the relevant literature suggested that participants not be told of information that has no direct clinical relevance. However, occupational studies differ from population-based studies in the sampling frame used and the types of intervention available. In occupational settings, "clinical relevance" could be defined as whether participants could take reasonable preventive or medical action based on the results. In the workplace, these reasonable actions could include various engineering, administrative, or behavioral controls [Weeks et al. 1991]. Clearly, where valid risks to workers are found in studies, notification is warranted.

7.3 Inherited Genetic Factors: Practice

7.3.1 Prevention and diagnosis

Genetic tests have been shown to be useful for various nonoccupational diseases in terms of disease diagnosis and individual risk assessment and provision of preventive services [Grody 2003; Burke et al. 2002]. Thus, they are becoming a part of general medical practice. The extent to which they will impinge on practice related to the workplace and workers is not known. Whether such approaches will be useful for occupational disease also is not known. If genetic tests are to be useful in occupational health, a process is needed so that evidence-based integration of data for the development of guidelines for disease prevention and health services occurs such as the guidelines that have been suggested for general clinical and public health practice [CDC 2007]. CDC established the Evaluation of Genomic Applications in Practice and Prevention (EGAPP) Working Group in 2005 to support the development of a systematic process to assess the available information regarding the validity and utility of emerging genetic tests for clinical practice. In 2007, EGAPP announced their first evidenced-based recommendation on the testing for cytochrome P450 polymorphisms in patients undergoing treatment for clinical depression using serotonin reuptake inhibitors. The EGAPP recommendation was to not conduct genetic testing at this time as the weight of evidence did not support the need for testing [EGAPP 2007]. Since the first evidenced based recommendation in 2007, five additional recommendations have been made.

In the future, the practice of occupational medicine may occur against the backdrop of individualized or personal medicine. At the least this may involve the need to consider an individual's genetic profile in the context of occupational exposures in terms of risk and prevention. The pressures to consider genetics and occupational exposures may grow as pharmacogenetic assessments become more common in medical practice [Rothstein 2003]. The question that arises is whether this information should be used in making workplace decisions.

Genetic tests for more than 1,400 clinical diseases are available, with approximately 300 more in the research and development stage [GeneTests 2009]. Laboratories performing genetic testing receive oversight under the Clinical Laboratory Improvement Amendment of 1988 (CLIA). CLIA was enacted to ensure and improve the accuracy and reliability of medical testing. It imposes basic requirements that address personnel qualifications, quality control and assurance, and degree of skill to perform and interpret. Specialty areas have been identified for which targeted requirements are determined. To date, no specialty area for genetic testing has been formed to tailor the requirements for genetic testing laboratories for the explosion of new tests and technologies currently in use [Javitt 2006]. One aspect of CLIA is proficiency testing. Only a few organizations offer proficiency testing for genetic tests and then only for a few of the genetic tests available [Javitt and Hudson 2006]. Genetic testing laboratories are left to determine their proficiency for themselves. It has been reported that for those laboratories that do perform proficiency testing also report fewer deficiencies and thus fewer analytical errors [Javitt and Hudson 2006]. In addition, research laboratories are exempt from CLIA regulations provided they do not give individual results for the purposed of diagnosis, treatment or prevention of diseases [Renegar et al. 2006].

The FDA does not generally regulate in-house developed tests. A few free-standing test kits are available and the FDA does review the clinical analytical validity and labeling claims [Javitt 2006; Javitt and Hudson 2006]. FDA also regulates a small subset of genetic tests known as in vitro diagnostic multivariate index assays. However, the majority of genetic tests are not affected by this regulation [Javitt 2006; Javitt and Hudson 2006] CLIA does not explicitly specify how accurate in-house tests need to be. Recently, several companies have begun offering genomic scans [Hunter et al. 2008]. These genomic scans do not fall under FDA or CLIA oversight, since the companies make the disclaimer that results should not be used for making medical decisions. In addition, the analytical validity, clinical validity and clinical utility have not been determined for these genomic scans [Hunter et al. 2008]. The proliferation of direct to consumer testing runs the risk of misleading consumers by providing inaccurate results thus undermining consumer confidence in genetic testing [Javitt and Hudson 2006].

7.3.2 Genetic screening and job actions

The capacity of the human body to respond to chemical exposure and physical agents varies from one individual to another. To some extent this is due to genetic characteristics which, in principle, could become part of employment testing known as genetic screening. Genetic screening is the examination of the genetic makeup of employees or job applicants for certain inherited traits. The actual use of genetic assays or tests of workers in job offering or placement is believed to be rare, but the available data to assess such activity are weak

[AMA 2004]. However, with the passing of GINA, the use of genetic information in employment decision is prohibited. Still it is useful to reflect on the concerns about using genetic information in work-related actions. This will be discussed further in this chapter. In 2009, the EEOC published proposed regulations regarding GINA's employment provisions. Title I and II of GINA establish legal protections from discrimination based on genetic information. Title II focuses on the workplace and prohibits employers, unions, employment agencies, and labor-management training programs from using genetic information in connection with employment decisions, bars intentional collection of genetic information regarding job applicants and employees, imposes confidentiality and record keeping requirements and prohibits retaliation [FRN 2009]. These regulations are expected to be effective November 21, 2009.

The respective roles of genetic and environmental factors in disease differ greatly [Grassman et al. 1998; Brain et al. 1988]. The certainty with which genetic characteristics can be used to predict a disease with or without an occupational exposure also varies widely [Lemmens 1997]. Determining the correct mix of genetic and environmental factors and the attendant risks is a complex endeavor fraught with many uncertainties. Therefore, the ability of employers or workers to make informed employment-related decisions based on genetic and environmental factors is limited by the degree of certainty about the relative roles of these factors. Predictions of risk that are highly uncertain will undermine claims of rights to freedom and well-being.

The degree of certainty in genetic screening has ethical implications. If employers or workers make employment decisions on the basis of tests with low predictive value, workers may be harmed or resources may be wasted. Genetic screening is less useful if the screening occurs after the exposure rather than before the worker begins the job and has exposure. The extent to which occupational safety and health investigators and practitioners are certain about the meaning of genetic information will influence the nature of communications to workers, employers, and others.

Another issue is that in today's workforce, workers are not always employees of the companies, but rather are contract, temporary, or subcontract workers. These workers do not have the same benefits and may be subject to differing philosophies regarding genetic information, testing, and the ramifications of that information.

The interpretation of information about a potential worker's health risk was seen in the case of *Echazabal* v. *Chevron* when the Supreme Court ruled that employers do not have to hire a person with a disability (in this case hepatitis C) if they believe the person's health or safety would be put at risk by performing the job [NCD 2003]. In this case, the job involved working around chemicals in a refinery. The case illustrates the potential for discrimination against employees who might be identified at increased risk through genetic screening or from genetic information in their medical records [Kim 2002].

Four objectives of genetic screening have been identified: (1) to ensure appropriate placement at the jobsite, (2) to exclude job applicants with increased susceptibility to disease, (3) to set limit values for more susceptible subgroups, and (4) to provide individual health counseling [Van Damme et al. 1995]. In general, pre- and postemployment nongenetic testing is a relatively common practice in selection and placement in the workplace. Susceptibility, however, is the result of a variety of genetic and nongenetic factors. Despite the profound advances in understanding the human genome, there are still no genetic tests that have been fully validated for use to screen perspective employees for occupational disease risks. Moreover, much controversy surrounds the practice of genetic screening, including such issues as the poor predictive value of the tests [Holtzman 2003; Van Damme et al. 1995]. Genetic polymorphisms may be unevenly distributed in the population among different ethnic groups [Rebbeck and Sankar 2005]. Thus, racial or ethnic discrimination could be a consequence of inappropriate use of genetic screening, which might be aimed at excluding workers at employment examination [Van Damme et al. 1995]. In the practice of occupational medicine, genetic information has been used selectively, mostly as derived from medical history, in job placement or diagnosis [AMA 2004, 1999; Staley 2003].

Of special concern for researchers and practitioners is how to communicate genetic screening results. Condit et al. [2000] suggested three goals for communication about genetics: (1) the focus should be on the health and well-being of the individual, (2) individual rights for free choice should be actively protected, and (3) stigmatization associated with specific genetic characteristics should be avoided.

The ethical issues of genetic screening of job applicants or workers have been assessed from various points of view. Genetic screening bears on the fundamental interests of workers, employers, and society [Gewirth 1998]. Various ethical issues of genetic screening have been addressed in the literature, including the certainty of the interpretation of genetic information, autonomy, privacy, confidentiality, discrimination, and stigmatization. Some authors argue against genetic screening in the workplace because of ethical concerns. Others argue that screening is supportable with appropriate safeguards [MacDonald and Williams-Jones 2002; Maltby 2000; Bingham 1998; Van Damme et al. 1995; Ashford et al. 1990; Gewirth 1986; Murray 1983]. In a review of the use of genetic information in the workplace, it was found that genetic screening should only be done with the consent of the worker with the worker controlling access to that information [Geppert et al. 2005]. In addition, genetic testing should only be done when the information was required to protect the safety of that worker or other workers.

Genetic screening has been assessed by the European Group on Ethics in Science and New Technologies [EGE 2003], which concluded that the use of genetic screening in the context of the medical examination, as well as the disclosure of results of previous genetic tests, is not ethically acceptable. Furthermore, EGE found that, to date, there is no proven evidence that the existing genetic screening tests have relevance or reliability in the context of employment. Generally, genetic screening tests still have uncertain predictive value [EGE 2003].

Most of the research to date has focused on just a few genetic polymorphisms. Workplace decisions based on such research would not yet be scientifically supportable, since few of these polymorphisms have been found to be definitive causative factors for occupational diseases. The attributable risk or proportion of risk as a result of a susceptibility-conferring genotype generally only reaches a level above 25% when the relative risk is about 5 and the frequency of the genotype is 10% or greater [Holtzman and Marteau 2000]. Previous studies of genetic polymorphisms have generally failed to identify groups of individuals with a relative risk greater than 3 [Holtzman and Marteau 2000]. One of the few exceptions found to date is the beryllium example, where a ninefold difference was found in risk of chronic beryllium disease for those who carried the *HLA-DPB1^{E69}* marker, which is present in about 40% of the population [McCanlies et al. 2002]. Hseih et al. [2007] reported a 13-fold increased risk of liver fibrosis among vinyl chloride exposed workers with *CYP2E1 c2c2* genotype. The tendency to reduce complex biological and social occupational phenomena to a single genetic cause and the small attributable risks that have been assessed reduce the confidence that most polymorphisms studied thus far would be defensible for genetic screening [Vineis et al. 2001].

Genetic screening information may be useful to inform potential employees of job risks if that information is not available to employers in individually identifiable form. While, in principle, it seems useful that prospective employees would benefit from information about potential risks, the attendant problems are not without impact. Using such a test, many false positive findings could occur resulting in people making employment decisions based on flawed information. Second, the difference between voluntary anonymous screening and mandated screening of individually identifiable applicants by prospective employers is huge with regards to what is known and not known about the relevance of genetic testing for occupational diseases. In contrast, if the test had a high (>90%) predictive value, would an employer have an argument for the right to use it in employee selections? An American Management Association survey of medical testing in the workplace conducted in 2004 found that some companies have used susceptibility to workplace hazards as a reason to hire, assign or reassign, dismiss, or retain employees [AMA 2004]. However, the employee may also benefit from the knowledge of genetic tests results by deciding whether to take a job or stay in a job. Decisions can be made regarding health or the need for increased medical monitoring.

7.3.3 Autonomy (self-determination), privacy, and confidentiality

Some observers believe that the central ethical question is whether using information obtained in genetic testing violates the rights of current or prospective workers [Bingham 1998]. In this view, genetic screening interferes with an individual's right to self-determination regarding employment if the employer uses genetic screening information for employment-related decisions. A small survey of workers found that workers had a strong interest in learning about personal genetic information and also felt that it needed greater protection, because of concerns about misuse, than general medical information [Roberts et al. 2005]. Safeguards have been called for to protect against the release and misuse of genetic information. It has been argued that genetic privacy has intrinsic value as a facet of autonomy and that respect for autonomy implies a duty to respect the genetic privacy of others [Anderlik and Rothstein 2001]. However, a different view holds that employers have interests (or responsibilities) to protect the well-being of their workers by using genetic screening information to select and place workers [French 2002; Krumm 2002; Anderlik and Rothstein 2001]. Employers have used this justification to support medical screening in general. Two federal class action lawsuits have been brought to the EEOC with regard to preplacement nerve tests to exclude workers with abnormal results. Although nerve testing is not a genetic test, the premise is the same. The EEOC lost both of these cases, so in essence, the practice is currently legal under federal law (*EEOC v. Rockwell International Corp.*, 60 F. Supp 2d 791 (N.D. Ill. 1999), aff'd, 243 F.3d 1012 (7th Cir. 2001); and *EEOC v. Woodbridge Corporation*, 124 F. Supp 2d 1132 (W.D. Mo. 2000), aff'd, 263 F.3d 812 (8th Cir. 2001)) [French 2002].

7.3.4 Stigmatization and discrimination

Identifying prospective or current workers with genetic risk factors may have a psychological and economic impact on individuals and groups [Marteau and Richards 1996; Billings et al. 1992]. Workers who are labeled as having an undesirable genetic trait may also have the mistaken impression that this trait puts them at risk from many or all exposures. Racial or ethnic groups who may already be burdened by discriminatory practices may be further burdened if they appear to have an inordinate frequency of various traits [Wiesner 1997].

Potential abuses and untoward effects of genetic screening are a concern. Genetic screening involves obtaining DNA, usually from a blood specimen, followed by the analysis of the DNA for genetic sequences, variants, polymorphisms, mutations, and deletions. Genetic screening creates information about a person that indicates, or appears to indicate, either the possibility of health risks from workplace exposures or the possibility of health effects unrelated to work. The findings of such tests are generally reported as a probability or possibility of occurrence.

Depending on who possesses genetic information and how they act, the potential abuses and untoward effects of genetic information about workers can include discrimination in employment and health and life insurance, labeling, individual and group stigmatization, and family disruption. While little evidence exists to support these contentions, many professional organizations and authoritative committees give strong credence to the possibility of these effects and advocate safeguards to protect against them [Watson and Greene 2001; U.S. Task Force on Genetic Testing 1998; ASHG 1996; ACOEM 1995]. A government-appointed committee in the United Kingdom did approve the use of genetic test results for Huntington's disease by British life insurance providers [Aldred 2000]. A U.S. health insurance provider has recommended its own guidelines for accessing genetic tests that include not using genetic test results to classify groups for the purpose of providing health coverage [Aetna 2002]. The general consensus is that at the present time the social consequences of revealing genetic screening test results may outweigh the benefits of valid and meaningful tests.

7.4 Inherited Genetic Factors: Litigation and Regulation

7.4.1 Litigation

One of the first workplace areas where genetic information has been used is workers' compensation. Even though traditional workers' compensation is a no-fault system under which an employer takes a worker "as he or she is," there is still potential for ethical and legal issues to arise involving genetic information. In the United States, there is no legal prohibition against including any medical or genetic tests in the independent medical examination that is routine in workers' compensation cases [Rothenberg et al. 1997]. In addition, informed consent for such testing is not required. By extension, genetic information may also be used as proof of causation in toxic injury litigation. However, "analysis of the role of genetic factors in multiple cause cases requires statistical and mechanistic data about how the genetic and toxic risks combine to cause disease" [Poulter 2001].

One example of the use of genetic information in a fault-based workers' compensation case involved genetic factors linked to occupational carpal tunnel syndrome [Schulte and Lomax 2003]. Railroad workers who filed for compensation under the Federal Employees Liability Act were tested for a genetic characteristic believed to predispose them to carpal tunnel syndrome. However, the genetic test had not been validated for this use. Second, unresolved is the question of whether society should use genetic testing for a susceptibility genotype to apportion causation. This question raises the issue of whether immutable traits beyond a worker's control should be factored into a claim of work-relatedness of a disease. In the above case, if it was found that genetic predisposition was a factor in the disease and not just occupational exposure, then the liability would have been reduced. The EEOC filed suit challenging the use of genetic testing in this case. In the settlement of *EEOC v. Burlington North Santa Fe Railway Co.* (CA 02.C0456) Burlington North was required to suspend genetic testing, could not analyze the results from any previous genetic testing, and could not analyze any blood specimens previously obtained [EEOC 2001].

Indeed in some jurisdictions (various states such as Iowa, Wisconsin, New York, and New Hampshire), consensual genetic testing is allowed in compensation cases. In the United States, most workers' compensation statutes permit medical testing, including genetic testing, to ascertain the medical condition of the claimant and the potential work-relatedness of the claim [Schulte 2004]. However, various U.S. organizations do not generally condone genetic testing without informed consent [ACOEM 2005].

Genetic information is also likely to be used in toxic tort lawsuits by both plaintiffs and defendants [Marchant 2003b]. In such cases where courts require that plaintiffs prove the defendant's actions double the background risks (i.e., relative risk greater than 2.0) to satisfy the "more likely than not" standard of causation, genetic information could be used to segment most populations and identify a subgroup with a particular genetic polymorphism with a relative risk greater than 2.

7.4.2 Differences in genetic susceptibilities and inclusion in risk assessments

Risk assessments are conducted to help decision-makers to determine the risk of exposure and risks to health. Despite familiar examples of interindividual variability, as well as emerging advances from molecular genetics, the potential application of such information to risk assessment has rarely been attempted [NRC 2007; Malaspina 1998]. Few examples exist of the incorporation of genetic information in quantitative risk assessments. The OSH Act stipulates that no worker should suffer impairment from work; however, occupational standards are clearly set at levels that include residual risks. Hypersusceptibility has never been a major factor in determining permissible exposure levels [Hornig 1988].

In risk assessment, genetic information may replace default assumptions when specific information regarding exposure, absorption, toxicokinetics, and species extrapolation is unavailable or limited [Ponce et al. 1998; Marchant 2003b]. Although examples of how genetic biomarker information can be used in risk assessments are limited [Dourson et al. 2005; Toyoshiba et al. 2004; El-Masri et al. 1999; Ponce et al. 1998; Hattis 1998; Bois et al. 1995; Hattis and Silver 1993], a general framework can be adduced [NRC 2007]. Genetic biomarkers can be used to stratify risks and identify high-risk subgroups. They also can be used to develop mechanism-based models for risk assessment [Toyoshiba et al. 2004].

Various technical questions abound about the use of genetic factors in risk assessment and need to be considered. For instance, does the tenfold uncertainty factor traditionally used to account for interindividual variability within the human population adequately describe the observed variability of response and susceptibility? Does the conservative default assumption used in cancer risk assessment account for interindividual variability relative to human exposure to carcinogens?

Genetic information has been used in risk assessment models to determine the impact of the role of metabolic polymorphisms on risk estimates [El-Masri et al. 1999; Bois et al. 1995]. While genetic information has the promise of more refined risk assessments through identification of gene-gene and gene-environment interactions, there is danger that various ethical and social issues will arise. These include stigmatization, discrimination, and the interpretation that removing a susceptible person from the exposure scenario without reducing exposure opportunities will reduce risk effectively, when it may not on a comparative basis [Holtzman 2003; Vineis et al. 2001; Ashford et al. 1990].

7.4.3 Regulation

Regarding the use of genetic information in occupational safety and health regulations, there are no examples of where such information is required. Genetic advances push at the historic boundaries of the OSH Act. The act mandates standards and rules to assure "to the extent feasible...that no employee will suffer material impairment of health or functional capacity." This raises the question of whether workers who could be defined by certain genetic polymorphisms as "hypersusceptible" should have special protections [Marchant 2003b]. The implementation of these protections raises a host of questions and issues regarding privacy, discrimination, and responsibility [Bergeson 2003]. Will employers have a duty to warn individuals with a genetic susceptibility for a specific workplace exposure? (See Marchant [2003b] for a more detailed discussion of these issues.)

Some genetic testing laboratories are using "direct to the consumer" advertisement to publicize their testing capabilities [Hunter et al. 2008; Javitt and Hudson 2006]. These 40 plus companies offer a variety of services such as being able to determine your ethnicity, to determining which sports your child may be better at to a dating service based on your genes. Most companies do not offer testing directly to the public. The underlying criticism is that consumers may be misled and not able to fully and correctly interpret the results and understand all of the implications with regard to their health. Access to clinical genetic testing is currently regulated by state laws. Some states allow consumer-requested genetic testing, while other states prohibit any medical testing unless requested by a physician. These "direct to consumer" companies by-pass state laws as they make no claims about their use for medical treatment.

7.5 Acquired Genetic Effects: Research

There is an extensive scientific literature assessing the impact of environmental hazards on genetic material [Hagmar et al. 2004; Bonassi et al. 2004; Jones et al. 2002; Albertini 2001; Toraason et al. 2001]. For the most part, this has involved assessment of cytogenetic effects (e.g., effects on chromosomes) and changes in various reporter genes such as GPA and HPRT, mutations, and the formation of DNA and protein adducts following exposure to electrophilic chemicals or ionizing radiation [Perera et al. 2003; Kelada et al. 2003; Phillips 2002; Albertini 2001; Groopman and Kensler 1999; Vineis et al. 1990; Ehrenberg et al. 1974]. The objectives of much of this research were to determine if genetic damage did occur and if it could lead to harmful health effects [Bonassi et al. 2004].

Much of the newer DNA and expression technologies, including toxicogenomics, transcriptomics, proteomics, and metabonomics, are means to assess acquired genetic effects [NRC 2007; Wang Z et al. 2005; Toyoshiba et al. 2004; Waters et al. 2003; Christiani et al. 2001]. These approaches allow for assessing the expression of many thousands of genes before and after exposure. Implicit in these approaches is that effects of xenobiotics can be detected in expression of genes. Critical in using this technology will be bioinformatics, the ability to analyze and interpret the vast amounts of data that arise from the studies. Such interpretation is quite difficult because many factors affect gene expression, and there is need to distinguish adaptive or homeostatic responses from pathologic ones. If a pattern from high throughput (e.g., microarrays) can be validated as a biomarker of effect, it may be used as an independent or dependent variable in etiologic or intervention research and as evidence of harm in workers' compensation or tort litigation [Segal et al. 2005; Marchant 2003a, 2003b]. These patterns could also be used in standards as biological exposure indices.

The informed consent issues raised in section 7.2 for research involving banked specimens and inherited genetic factors also pertain to research on acquired genetic effects. Proteomic, toxicogenomic, and transcriptomic research may occur with specimen banks previously collected. Individuals who participate in such banks may only be able to give broad, general consent. Pertinent to this discussion are issues of privacy and confidentiality in the use of banked specimens in research [Rothstein and Knoppers 2005].

7.6 Acquired Genetic Effects: Practice

The ascertainment of acquired genetic damage information in occupational safety and health practice would generally occur in the form of genetic monitoring. However, the fact that it involves preclinical somatic genetic effects often leads to its consideration as a somewhat different form of monitoring. Genetic monitoring is similar to biological monitoring, but instead of merely assessing exposure, it assesses the effects of exposure. At present, the results of genetic monitoring can only be interpreted on a group level; they have not been validated as individual risk predictors [Van Damme et al. 1995]. If high-throughput expression technologies become candidates for use in genetic monitoring, the issues of standardization, validation, and interpretability will have to be overcome since these will be much greater than with a single test.

7.7 Acquired Genetic Effects: Regulation and Litigation

Currently, no U.S. regulations require genetic monitoring of workers. In part, this is because questions arise about whether genetic monitoring indicates exposure, a potential health problem, or a compensable injury [Schulte and DeBord 2000].

The gene expression technologies have been viewed as potentially providing useful data for group risk assessment; however, there are numerous interpretive questions, as summarized by Freeman [2004], regarding the use of data from microarray experiments by regulating agencies.

- How does a regulator deal with risk assessment data that scientists are often unable to interpret—data that some companies are anxious to submit and others to withhold?
- How does this same regulator evaluate information that is produced without a universally recognized standard for laboratory protocols or data formats?
- Should companies submit all data voluntarily without knowing whether regulators will be able to understand it, and, if so, exactly how they will use it?
- What if data that cannot be interpreted now are later shown to indicate toxicity, perhaps at a low level that could not be detected in animal testing [Freeman 2004]? The critical issue in using genomics data is that if and when it is interpretable in terms of population risks, what will be the regulatory focus if sensitive subgroups are identified? Will controls be required to protect these groups, or will risk management strategies, such as communications, be applied [Freeman 2004]?

Data from gene expression technologies may contribute to understanding the impact of interindividual variability in risk assessments. Hattis [1998] has described examples that show various possibilities for improving quantitative risk assessment of both cancer and noncancer effects of environmental and occupational exposures with the aid of human data on interindividual variability. He argues that improvements would be possible if interindividual variability data were collected more systematically by investigators and if previously collected individual data were made more readily available. In general, there appears to be a strong consensus among risk assessors that reducing the uncertainty associated with our understanding of human variability will improve risk assessment [Hattis and Swedis 2001; Bailar and Bailer 1999; Grassman et al. 1998].

In the short term, transcriptomics, proteomics, and metabonomics will probably be of most value for the hazard identification aspect of risk assessment [Morgan et al. 2002; Faustman and Omenn 1996]. If gene expression technology is to enter the mainstream of the risk assessment process, protocols for assays to confirm selected biochemical responses will need to be developed as regulatory requirements [Morgan et al. 2002]. Various uncertainties exists that limit confidence in and utility of risk assessment in general for informing regulatory decisions. These include issues in extrapolating from animals to humans, high to low doses, the shape of the dose-response curve, and estimating workplace levels of exposure. Genetic expression data may be useful in addressing these issues [Marchant 2003a].

Genetic expression data potentially may be used to quantify or address exposure in litigation [Marchant 2003a]. Critical in this regard is the timing of these expression changes, the linkage to a particular exposure, recovery for latent risks, and decisions on interventions such as medical monitoring [Marchant 2003a]. The key feature for any of these uses is that the genetic expression tests be validated for the specific use being considered.

One potential outcome of genetics research in general, and occupational toxicogenomics research in particular, is the potential to transform current conceptions of "risk" and "injury" in the law of toxic torts [Grodsky 2007]. As more is learned about acquired genetic effects, such as those related to gene expression, preclinical changes will be detected before classical clinical symptoms occur. The question has been raised whether plaintiffs exposed to toxic hazards and placed at significant risk of disease, yet perhaps not physically "injured," should be entitled to some form of legal remedy [Grodsky 2007]. This contention hinges on the evidence base regarding the validity of preclinical change to predict or lead to disease. Blurring of the risk and injury concepts may lead to ambiguity about the appropriate public health action or legal remedies. Where science can not only identify such conditions but possibly treat them at the molecular level, a case can be made for the need for medical monitoring and possibly treatment of individuals to prevent or minimize the ultimate effect. This concept is not new to the occupational safety and health field, but has been advocated for workers determined by epidemiologic research to be members of high-risk groups, such as asbestos workers [Samuels 1998b]. This was termed "high-risk management." If new genetic and genomic technologies and research enable the identification of an expanding progression of biologic effects between a chemical (or other workplace) exposure and fully developed disease, there is the expectation of new legal claims by such workers [Grodsky 2007]. The occupational safety and health community may have to apply high-risk management concepts to these groups of workers. This would mean considering appropriate risk communication, biological monitoring, medical screening, and risk management issues [Schulte 2005; Samuels 1998b].

Another way to communicate results to workers is through group presentations that stress the overall results and how they fit in with current research knowledge. Such presentations provide a forum for questions and discussions about the issues, as well as education about interpretation of results. In the beryllium example, some workers with chronic beryllium disease do not have the $HLA\text{-}DPB1^{E69}$ gene, and it is important to emphasize to workers that absence of the high-risk marker does not mean absence of risk for disease. In other words, good workplace hygiene is still essential. Care also must be taken to ensure that workers are not inadvertently identified when presenting group results through too much detail in tables and examples.

7.8 The Adequacy of Safeguards to Protect Workers Against Misuse of Genetic Information

Rothstein [2000b] reviewed the laws related to genetic nondiscrimination.

As of October 2000, about half of the states have enacted laws prohibiting genetic discrimination in employment. President Clinton signed a similar Executive Order applicable to federal employees on February 10, 2000 (Executive Order No. 13145) [65 Fed. Reg. 6877 (2000)]. The laws are directed at two perceived problems: (1) employers responsible for employee and dependent medical expenses, either through commercial insurance or self-insurance, have a great economic incentive to exclude presumed high-cost future consumers of medical resources [EEOC 1995] and (2) individuals who are at a genetic risk of disease will be discouraged from undergoing genetic testing if they think that their current or future employers would have access to the results of the tests or other genetic information [Yesley 1999].

Regarding safeguards in genetic research, Anderlik and Rothstein [2001] presented the following discussion:

> *Although the rules for the new science are not yet fixed, we find some areas of consensus. First, privacy is too large an issue to be solely the responsibility of geneticists, or any other group. The involvement of ethicists, social scientists, lawyers, and representatives of affected communities in appropriate cases is an important protection against the errors of judgment that may result from narrowness of perspective. Without public participation, there is a very real risk that the public will turn against genetic research if projects come to light that violate public expectations of protection of privacy and autonomy.*
>
> *Second, individuals and organizations working in the field of genetics should add privacy protection to the checklist of items to be reviewed at each stage of a project, from conception through ongoing monitoring.*
>
> *Third, although laws protecting the privacy of health information and prohibiting genetic discrimination are in place in most jurisdictions, there are gaps in these laws and in the social safety net. Public fears of irrational and rational discrimination in insurance are not unjustified, and scientists eager to recruit participants for genetic research will have to address these fears.*

Researchers are not regulated directly unless they fit the HIPAA definition of health care providers. In addition, the key provision of HIPAA affecting research permits covered entities (such as health care providers) to use or disclose protected health information for research purposes without authorization by the research participant. There are, however, limited safeguards for research participants. Before information can be used or disclosed in this manner, the covered entity must obtain a written waiver of authorization from an IRB or a privacy board for each research protocol.

This requirement applies to all research for which a covered entity serves as a conduit of protected health information, regardless of funding source.

The federal government offers some additional confidentiality protection for federally funded research under 301(d) and 308(d) of the Public Health Service Act [FDA 2007]. Assurance under 308(d) protects both the individual and the institution, while a certificate of confidentiality under 301(d) protects only the individual in the research study. These protections limit disclosures of information that are permitted under the Privacy Act, such as routine or court-ordered disclosures without the consent of the respondent [NIH 2007]. These additional protections under 301(d) and 308(d) are generally reserved for data collection of sensitive information.

Unfortunately, as long as legal protections remain imperfect, one of the principal tasks for researchers committed to ethical conduct will be educating potential participants about the harms that may be associated with participation in genetic research. Before educating potential participants, however, genetic researchers need to consider the societal aspects of their research. For better or worse, privacy will be as important to genetic researchers as pedigrees, polymorphisms, and proteomics.

In summary, genetic information has the potential to improve employee health and reduce worker disability. State and Federal laws have been passed to protect individuals against using genetic information for discriminatory practices. Some concerns still remain with regard to autonomy, privacy, stigmatization and clinical relevance. At this time, genetic screening in the workplace is not recommended as currently no genetic test has been validated for an occupational disease.

References

AAOHN (American Association of Occupational Health Nurses, Inc.) [2004a]. Code of ethics. http://www.aaohn.org/practice/ethics.cfm. Date accessed: January 23, 2007.

AAOHN (American Association of Occupational Health Nurses, Inc.) [2004b]. Position statement: confidentiality of employee health information. http://www.aaohn.org/practice/positions/upload/Confidentiality_of_Emp_Health_Info.pdf. Date accessed: January 23, 2007.

Abdel-Rahman SZ, Ammenheuser MM, Omiecinski CJ, Wickliffe JK, Rosenblatt JI, Ward JB Jr. [2005]. Variability in human sensitivity to 1,3-butadiene: influence of polymorphisms in the 5'-flanking region of the microsomal epoxide hydrolase gene (*EPHX1*). Toxicol Sci *85*:624–631.

ACCE (analytical validity, clinical validity, clinical utility, and ethical, legal, and social implications and safeguards) [2007]. Evaluation of genetic testing http://www.cdc.gov/genomics/gtesting/ACCE.htm. Date accessed: May 11, 2007.

ACOEM (American College of Occupational and Environmental Medicine) [2005]. Genetic screening in the workplace. Approved October 27, 2005. ACOEM position statement: http://www.acoem.org/guidelines.aspx?id=726. Date accessed: February 4, 2008.

ACOEM (American College of Occupational and Environmental Medicine) [1995]. ACOEM position on the confidentiality of health information in the workplace. J Occup Environ Med *37*:594–596.

ADA (Americans With Disabilities Act) [1990]. http://www.ada.gov/pubs/ada.htm. Date accessed: March 24, 2003.

Aetna [2002]. Aetna recommends guidelines for access to genetic testing: company advocates education and information privacy. http://www.aetna.com/news/2002/pr_20020617.htm. Acessed April 23, 2006.

Albertini RJ, Sram RJ, Vacek PM, Lynch J, Nicklas JA, van Sittert NJ, Boogaard PJ, Henderson RF, Swenberg JA, Tates AD, Ward JB Jr., Wright M, Ammenheuser MM, Binkova B, Blackwell W, de Zwart FA, Krako D, Krone J, Megens H, Musilova P, Rajska G, Ranasinghe A, Rosenblatt JI, Rossner P, Rubes J, Sullivan L, Upton P, Zwinderman AH [2003]. Biomarkers in Czech workers exposed to 1,3-butadiene: a transitional epidemiologic study. Res Rep Health Eff Inst *116*:1–141.

Albertini RJ [2001]. Validated biomarker responses influence medical surveillance of individuals exposed to genotoxic agents. Radiat Prot Dosimetry *97*:47–54.

Albertini RJ, Hayes RB [1997]. Somatic cell mutations in cancer epidemiology. In: Applications of biomarkers in cancer epidemiology. IARC Scientific Publication No. 142. Lyon, France: International Agency for Research on Cancer, pp. 159–184.

Aldred C [2000]. Insurers to use genetic info. Bus Insur *34*:1–2.

Allen AL [1997]. Genetic privacy: emerging concepts and values. In: Rothstein MA, ed. Genetic secrets: protecting privacy and confidentiality in the genetic era. New Haven, CT: Yale University Press, pp. 31–59.

Altshuler D, Hirschhorn JN, Klannemark M, Lindgren CM, Vohl MC, Nemesh J, Lane CR, Schaffner SF, Bolk S, Brewer C, Tuomi T, Gaudet D, Hudson TJ, Daly M, Groop L, Lander ES [2000]. The common PPAR gamma Pro12Ala polymorphism is associated with decreased risk of type 2 diabetes. Nat Genet *26*:76–80.

AMA (American Management Association) [1999]. AMA Workplace testing survey: medical testing. Annual research report. New York: AMA, April 1999.

AMA (American Management Association) [2004]. Workplace testing: medical testing. Annual research report 2004. New York: AMA. September 3, 2003 (http://www.amanet.org/research/archive.htm). Date accessed: December 11, 2006.

American Medical Association Council on Ethical and Judicial Affairs [1991]. Use of genetic testing by employers. JAMA *266*:1827–1830.

Anderlik MR [2003]. Commercial biobanks and genetic research: ethical and legal issues. Am J Pharmacogenomics *3*:203–215.

Anderlik MR, Rothstein MA [2001]. Privacy and confidentiality of genetic information: what rules for the new science. Annu Rev Genomics Hum Genet *2*:401–433.

Andrews LB [1997]. Gen-etiquette: genetic information family relationships and adoption. In: Rothstein MA, ed. Genetic secrets: protecting privacy and confidentiality in the genetic era. New Haven, CT: Yale University Press, pp. 255–280.

Arlt VM, Glatt H, Gamboa da Costa G, Reynisson J, Takamura-Enya T, Phillips DH [2007]. Mutagenicity and DNA adduct formation by the urban air pollutant 2-nitrobenzanthrone. Toxicol Sci *98*:445-457.

ASCO (American Society of Clinical Oncology) [2003]. Policy statement update: genetic testing for cancer susceptibility. J Clin Oncol *21*:1–10.

Ashford NA, Spadafor C, Hattis D, Caldart CC eds [1990]. Monitoring the worker for exposure and disease: scientific, legal and ethical considerations in the use of biomarkers. In: Johns Hopkins series in environmental toxicology. Baltimore, MD: The Johns Hopkins University Press pp.1-244.

ASHG (American Society of Human Genetics) [2000]. Should family members about whom you collect only medical history information for your research be considered "human subjects?" http://ashg.org/pages/statement_32000.shtml. Date accessed: September 10, 2008.

ASHG (American Society of Human Genetics) [1996]. Statement on informed consent for genetic research. Am J Hum Genet *59*:471–474.

Bailar JC III, Bailer AJ [1999]. Common themes at the workshop on uncertainty in the risk assessment of environmental and occupational hazards. Ann N Y Acad Sci *895*:373–376.

Barker PE [2003]. Cancer biomarker validation: standards and process. Ann N Y Acad Sci *983*:142–150.

Barker RN, Burney PG, Devereux G, Hardiman M, Holt PG, O'Donnell M, Platts-Mills TA, Seaton A, Strachan DP, Weiss ST, Woodcock A [2003]. The increase in allergic disease: environment and susceptibility. Proceedings of a symposium held at the Royal Society of Edinburgh, 4th June 2002. Clin Exp Allergy *33*:394–406.

Barrett JC, Vainio H, Peakall D, Goldstein BD. [1997]. 12th Meeting of the scientific group on methodologies for the safety evaluation of chemicals: susceptibility to environmental hazards. Environ Health Perspect *105*(Suppl 4):699–737.

Bates S [2001]. Scientific friction: when advancements in genetics collide with privacy and discrimination concerns, sparks fly, fears increase and litigation ensues. HRMagazine *46*(7):1–3. http://www.shrm.org/hrmagazine/2001index/0701/0701cov.asp. Date accessed: July 7, 2008.

Battuello K, Furlong C, Fenske R, Austin MA, Burke W [2004]. Paraoxonase polymorphisms and susceptibility to organophosphate pesticides. In: Khoury MJ, Little J, Burke W, eds. Human genome epidemiology: a scientific foundation for using genetic information to improve health and prevent disease. 3rd ed. New York: Oxford University Press, pp. 305–321.

Beauchamp TL, Childress JE, eds. [1994]. Principles of biomedical ethics. 4th ed. New York: Oxford University Press, pp. 406–418.

Beaudet AL, Belmont JW [2008]. Array-based DNA diagnostics: let the revolution begin. Annu Rev Med *59*:113-129.

Becker KG, Barnes KC, Bright TJ, Wang A [2004]. The Genetic Association Database. Nat Genet *36*:431–432.

Belmont Report [1979]. http://www.hhs.gov/ohrp/humansubjects/guidance/belmont.htm. Date accessed: May 3, 2007.

Benhamou S, Sarasin A [2005]. ERCC2/XPD gene polymorphisms and lung cancer: a HuGE review. Am J Epidemiol *161*:1–14.

Berg K, ed. [1979]. Inherited variation in susceptibility and resistance to environmental agents. In: Genetic damage in man caused by environmental agents. San Diego, CA: Academic Press, pp. 1–25.

Bergeson LL [2003]. Toxicogenomics and the workplace: what you need to know. Manufacturing Today *Jan/Feb*:8–9.

Beskow LM, Burke W, Merz JF, Barr PA, Terry S, Penchaszadeh VB, Gostin LO, Gwinn M, Khoury MJ [2001]. Informed consent for population-based research involving genetics. JAMA *286*:2315–2321.

Bilban M, Jakopin CB, Ogrinc D [2005]. Cytogenetic tests performed on operating room personnel (the use of anaesthetic gases). Int Arch Occup Environ Health *78*:60–64.

Billings PR, Kohn MA, de Cuevas M, Beckwith J, Alper JS, Natowicz MR [1992]. Discrimination as a consequence of genetic testing. Am J Hum Genet *50*:476–482.

Bingham E [1998]. Ethical issues of genetic testing for workers. In: Mendelsohn ML, Mohr LC, Peeters JP, eds. Biomarkers: medical and workplace applications. Washington, DC: Joseph Henry Press, pp. 415–422.

Blakely WF, Prasanna PGS, Grace MB, Miller AC [2001]. Radiation exposure assessment using cytological and molecular biomarkers. Radiat Prot Dosimetry *97*:17–23.

Blanchard AP, Hood L [1996]. Sequence to array: probing the genome's secrets. Nat Biotechnol *14*:1649.

Boffetta P, van der Hel O, Norppa H, Fabianova E, Fucic A, Gundy S, Lazutka J, Cebulska-Wasilewska A, Puskailerova D, Znaor A, Kelecsenyi Z, Kurtinaitis J, Rachtan J, Forni A, Vermeulen R, Bonassi S [2007]. Chromosomal aberrations and cancer risk: Results of a cohort study from central Europe. Am J Epidemiol *165*:36–43.

Bois FY, Krowech G, Zeise L [1995]. Modeling human interindividual variability in metabolism and risk: the example of 4-aminobiphenyl. Risk Anal *15*:205–213.

Bonassi S, Abbondandolo A, Camurri L, Dal Pra L, De Ferrari M, Degrassi F, Forni A, Lamberti L, Lando C, Padovani P, Sbrana I, Vecchio D, Puntoni R [1995]. Are chromosome aberrations in circulating lymphocytes predictive of future cancer onset in humans? Preliminary results of an Italian cohort study. Cancer Genet Cytogenet *79*:133–135.

Bonassi S, Hagmar L, Strömberg U, Montagud AH, Tinnerberg H, Forni A, Heikkilä P, Wanders S, Wilhardt P, Hansteen I-L, Knudsen LE, Norppa H. [2000]. Chromosomal aberrations in lymphocytes predict human cancer independently of exposure to carcinogens. European study group on cytogenetic biomarkers and health. Cancer Res *60*:1619–1625.

Bonassi S, Znaor A, Ceppi M, Lando C, Chang WP, Holland N, Kirsch-Volders M, Zeiger E, Ban S, Barale R, Bigatti MP, Bolognesi C, Cebulska-Wasilewska A, Fabianova E, Fucic A, Hagmar L, Joksic G, Martelli A, Migliore L, Mirkova E, Scarfi MR, Zijno A, Norppa H, Fenech M [2007]. An increased micronucleus frequency in peripheral blood lymphocytes predicts the risk of cancer in humans. Carcinogenesis *28*:625–631.

Bonassi S, Znaor A, Norppa H, Hagmar L [2004]. Chromosomal aberrations and risk of cancer in humans: an epidemiological perspective. Cytogenet Genome Res *104*:376–382.

Borman S [1996]. DNA chips come of age. Chem Eng News, December 9, pp. 42–43.

Botto LD, Yang Q [2000]. *5,10-Methylenetetrahydrofolate reductase* gene variants and congenital anomalies: a HuGE review. Am J Epidemiol *151*:862–877.

Boysen G, Georgieva NI, Upton PB, Walker VE, Swenberg JA. [2007]. N-Terminal globin adducts as biomarkers for formation of butadiene derived epoxides. Chem Biol Interact *166*:84-92.

Brain JD, Beck B, Warren A, Shaikh A, eds. [1988]. Variations in susceptibility to inhaled pollutants: identification, mechanisms and policy implications. Baltimore, MD: Johns Hopkins University Press pp.1-502.

Brandt-Rauf PW, Brandt-Rauf SI [1997]. Biomarkers: scientific advances and social implications. In: Rothstein MA, ed. Genetic secrets: protecting privacy and confidentiality in the genetic era. New Haven, CT: Yale University Press, pp. 184–196.

Brazma A, Hingamp P, Quackenbush J, Sherlock G, Spellman P, Stoeckert C, Aach J, Ansorge W, Ball CA, Causton HC, Gaasterland T, Glenisson P, Holstege FC, Kim IF, Markowitz V, Matese JC, Parkinson H, Robinson A, Sarkans U, Schulze-Kremer S, Stewart J, Taylor R, Vilo J, Vingron M [2001]. Minimum information about a microarray experiment (MIAME)-toward standards for microarray data. Nat Genet *29*:365–371.

Brennan P, Lewis S, Hashibe M, Bell DA, Boffetta P, Bouchardy C, Caporaso N, Chen C, Coutelle C, Diehl SR, Hayes RB, Olshan AF, Schwartz SM, Sturgis EM, Wei Q, Zarras AI, Benhamou A [2004]. Pooled analysis of *alcohol dehydrogenase* genotypes and head and neck cancer: a HuGE review. Am J Epidemiol *159*:1–16.

Brockton N, Little J, Sharp L, Cotton SC [2000]. *N-Acetyltransferase* polymorphisms and colorectal cancer: a HuGE review. Am J Epidemiol *151*:846–861.

Brookes AJ, Chanock SJ, Hudson TJ, Peltonen L, Abecasis G, Kwok P-Y, Scherer SW [2009]. Genomic variation in a global village: report of the 10th annual human genome variation meeting 2008. Hum Mutat *30*:1–5.

Burke W, Atkins D, Gwinn M, Guttmacher A, Haddow J, Lau J, Palomaki G, Press N, Richards CS, Wideroff L, Wiesner GL [2002]. Genetic test evaluation: information needs of clinicians, policy makers, and the public. Am J Epidemiol *156*:311–318.

Burris S, Gostin LO, Tress D [2000]. Public health surveillance of genetic information: ethical and legal responses to social risk. In: Khoury MJ, Burke W, Thomson EJ, eds. Genetics and public health in the 21st century; using genetic information to improve health and prevent disease. New York: Oxford University Press, pp. 527–548.

Carlsten C, Sagoo GS, Frodsham AJ, Burke W, Higgins JP [2008]. Glutathione S-transferase M1 (GSTM1) polymorphisms and lung cancer: a literature-based systematic HuGE review and meta-analysis. Am J Epidemiol *167*:759-774.

Carreón T, LeMasters GK, Ruder AM, Schulte PA [2006a]. The genetic and environmental factors involved in benzidine metabolism and bladder carcinogenesis in exposed workers. Front Biosci *11*:2889–2902.

Carreón T, Ruder AM, Schulte PA, Hayes RB, Rothman N, Waters M, Grant DJ, Boissy R, Bell DA, Kadlubar FF, Hemstreet GP, Yin S, LeMasters GK [2006b]. *N-acetyltransferase* 2 slow acetylation and for bladder cancer in workers exposed to benzidine. Int J Cancer *118*:161–168.

Cartwright RA, Glashan RW, Rogers HJ, Ahmad RA, Barham-Hall D, Higgins E, Kahn MA [1982]. Role of *N-acetyltransferase* phenotypes in bladder carcinogenesis: a pharmacogenetic epidemiological approach to bladder cancer. Lancet *2*:842–845.

CDC (Centers for Disease Control and Prevention) [2007]. Genetic testing. [http://www.cdc.gov/genomics/gtesting.htm]. Date accessed: July 2007.

Celi KA, Akbas E [2005]. Evaluation of sister chromatid exchange and chromosomal aberration frequencies in peripheral blood lymphocytes of gasoline station attendants. Ecotoxicol Environ Saf *60*:106–112.

CFR [2007]. Code of Federal regulations. Washington, DC: U.S. Government Printing Office, Office of the Federal Register. http://www.gpoaccess.gov/cfr/index.html. Date accessed: March 4, 2009.

Chan-Yeung M, Ashley MJ, Corey P, Maledy H [1978]. Pi phenotypes and the prevalence of chest symptoms and lung function abnormalities in workers employed in dusty industries. Am Rev Respir Dis *117*:239–245.

Chen Y, Li G, Yin S, Xu J, Ji Z, Xiu X, Liu L, Ma D [2007]. Genetic polymorphisms involved in toxicant-metabolizing enzymes and the risk of chronic benzene poisoning in Chinese occupationally exposed populations. Xenobiotica *37*:103–112.

Cherry N, Mackness M, Durrington P, Povey A, Dippnall M, Smith T, Mackness B [2002]. *Paraoxonase (PON1)* polymorphisms in farmers attributing ill health to sheep dip. Lancet *359*:763–764.

Christiani DC, Sharp RR, Collman GW, Suk WH [2001]. Applying genomic technologies in environmental health research: challenges and opportunities. J Occup Environ Med *43*:526–533.

Clamp M, Fry B, Kamal M, Xie X, Cuff J, Lin MF, Kellis M, Lindblad-Toh K, Lander ES. [2007]. Distinguishing protein-coding and noncoding genes in the human genome. Proc Natl Acad Sci U S A. 104:19428-19433.

Clayton EW [2003]. Ethical, legal and social implications of genomic medicine. New Engl J Med *349*:562–569.

Collins FS, Green ED, Guttmacher AE, Guyer MS [2003]. U.S. National Genome Research Institute: a vision for the future of genomics research. Nature *422*:835–847.

Condit CM, Parrott RL, O'Grady B [2000]. Principles and practices of communication processes for genetics in public health. In: Khoury MJ, Burke W, Thomson EJ, eds. Genetics and public health in the 21st century: using genetic information to improve health and prevent disease. New York: Oxford University Press, pp. 549–567.

Constable S, Johnson MR, Pirmohamed M [2006]. Pharmacogenetics in clinical practice: considerations for testing. Expert Rev Mol Diagn *6*:193–205.

Cookson WO [2002]. Asthma genetics. Chest *121*:7S–13S.

Cote ML, Chen W, Smith DW, Benhamou S, Bouchardy C, Butkiewicz D, Fong KM, Gene M, Hirvonen A, Kiyohara C, Larsen JE, Lin P, Raaschou-Nielsen O, Povey AC, Reszka E, Risch A, Schneider J, Schwartz AG, Sorensen M, To-Figueras J, Tokudome S, Pu Y, Tang P, Wenzlaff AS, Wikman H, Taioli E [2009]. Meta- and pooled analysis of GSTP1 polymorphisms and lung cancer: a HuGE-GSEC review. Am J Epidemiol *169*:802–814.

Cotton SC, Sharp L, Little J, Brockton N [2000]. *Glutathione S transferase* polymorphisms and colorectal cancer: a HuGE review. Am J Epidemiol *151*:7–32.

Coughlin SS, Hall IJ [2002]. *Glutathione S-transferase* polymorphisms and risk of ovarian cancer: a HuGE review. Genet Med *4*:250–257.

Couzin J, Kaiser J [2007]. Closing the net on common disease genes. Science *316*:820–822.

Dahl R [2003]. The EGP at five years. Environ Health Perspect *111*:A573.

Davey Smith G [2001]. Reflections on the limitations to epidemiology. J Clin Epidemiol *54*:325–331.

Davey Smith G, Ebrahim S [2003]. 'Mendelian randomization': can genetic epidemiology contribute to understanding environmental determinants of disease? Int J Epidemiol *32*(1):1–22.

D'Errico A, Taioli E, Chen X, Vineis P [1996]. Genetic metabolic polymorphisms and the risk of cancer: a review of the literature. Biomarkers *1*:149–173.

DeCoster S, Koppen G, Bracke M, Schroijen C, Den Hond E, Nelen V, Van de Mieroop E, Bruckers L, Bilau M, Baeyens W, Schoeters G, van Larebeke N [2008]. Pollutant effects on genotoxic parameters and tumor-associated protein levels in adults: a cross sectional study. Environ Health *7*:26.

Deng H, Zhang M, He J, Wu W, Zheng W, Lou J, Wang B [2005]. Investigating genetic damage in workers occupationally exposed to methotrexate using three genetic end-points. Mutagenesis *20*:351-357.

DHHS (Department of Health and Human Services) [2007]. Security and Privacy 45 CFR Part 160, 164. http://www.access.gpo.gov/nara/cfr/waisidx_07/45cfrv1_07.html. Date accessed: April 23, 2009.

DHHS (U.S. Department of Health and Human Services) [2005]. HHS protection of human subjects. 45 CFR Part 46. http://www.hhs.gov/ohrp/humansubjects/guidance/45cfr46.htm. Date accessed: April 23, 2009.

Diamandis EP [2006]. Peptidomics for cancer diagnosis: present and future. J Proteome Res *5*:2079–2082.

Dianzani I, Gibello L, Biava A, Giordano M, Bertolotti M, Betti M, Ferrante D, Guarrera S, Betta GP, Mirabelli D, Matullo G, Magnani C [2006]. Polymorphisms in DNA repair genes as risk factors for asbestos-related malignant mesothelioma in a general population study. Mutat Res *599*:124–134.

DOE (U.S. Department of Energy) [2001]. Human genome program, human genome management information system, genome glossary. http://www.ornl.gov/TechResources/HumanGenome/glossary/glossary.html. Date accessed: January 16, 2009.

Dorman JS, Bunker CH [2000]. *HLA-DQ* locus of the human leukocyte antigen complex and type 1 diabetes mellitus: a HuGE review. Epidemiol Rev *22*:218–227.

Dourson M, Haber LT, Maier A, Clewell HJ III, Hacks CE, Gentry PR, Covington TR [2005] Considering genetic differences in risk assessment. LRI Perspectives March. http://www.americanchemistry.com/s_acc/bin.asp?CID=1405&DID=5147. Date accessed: November 5, 2007.

Drug Abuse Prevention, Treatment, and Rehabilitation Act [2007]. http://ecfr.gpoaccess.gov/cgi/t/text/text-idx?c=ecfr&tpl=/ecfrbrowse/Title42/42cfr2_main_02.tpl. Date accessed: June 5, 2008.

EBP (Exposure Biology Program) [2007]. The GEI Initiative: Exposure Biology Program. http://www.gei.nih.gov/exposurebiology/index.asp. Date accessed: February 20, 2009.

EEOC (Equal Employment Opportunity Commission) [2001]. EEOC settles ADA suit against BNSF for genetic bias. April 18, 2001. http://www.eeoc.gov/press/4-18-01.html. Date accessed: February 20, 2008.

EEOC (Equal Employment Opportunity Commission) [2000]. EEOC policy guidance on executive order 13145: to prohibit discrimination in federal employment based on genetic information. Notice No. 915.002, July 26, 2000. http://www.eeoc.gov/policy.docs/guidance-genetic.html. Date accessed: February 20, 2008.

EEOC (Equal Employment Opportunity Commission) [1995]. Compliance manual, 902–45 (March 14, 1995), http://www.eeoc.gov/policy/compliance.html Reprinted in Daily Lab., Rep March 16 at E–1, E–23.

EGE (The European Group on Ethics in Science and New Technologies to the European Commission) [2003]. Opinion on the ethical aspects of genetic testing in the workplace. Opinion No. 18. http://www.ec.europa.eu/european_group_ethics/archive/2001_2005/avis_en.htm, Date accessed: October 19, 2006.

EGAPP (Evaluation of Genomic Applications in Practice and Prevention Working Group) [2007]. Recommendations from the EGAPP Working Group: testing for cytochrome P450 polymorphisms in adults with nonpsychotic depression treated with selective serotonin reuptake inhibitors. Genet Med 9:819-825.

EGP (Environmental Genome Project) [2008]. http://www.niehs.nih.gov/research/supported/programs/egp/. Date accessed: November 28, 2008.

Ehrenberg L, Hiesche KD, Osterman-Golkar S, Wenneberg I [1974]. Evaluation of genetic risks of alkylating agents: tissue doses in the mouse from air contaminated with ethylene oxide. Mutat Res *24*:83–103.

Eichner JE, Dunn ST, Perveen G, Thompson DM, Stewart KE, Stroehla BC [2002]. *Apolipoprotein E* polymorphism and cardiovascular disease: a HuGE review. Am J Epidemiol *155*:487–495.

Elbaz A, Levecque C, Clavel J, Vidal J-S, Richard F, Amouyel P, Alperovitch A, Chartier-Harlin M-C, Tzourio C [2004]. *CYP2D6* polymorphism, pesticide exposure, and Parkinson's disease. Ann Neurol 55:430–434.

Ellenbecker MJ [1996]. Engineering controls as an intervention to reduce worker exposure. Am J Ind Med *29*:303–307.

El-Masri HA, Bell DA, Portier CJ [1999]. Effects of *glutathione transferase theta* polymorphism on the risk estimates of dichloromethane to humans. Toxicol Appl Pharmacol *158*:221–230.

Engel LS, Taioli E, Pfeiffer R, Garcia-Closas M, Marcus PM, Lan Q, Boffetta P, Vineis P, Autrup H, Bell DA, Branch RA, Brockmoller J, Daly AK, Heckbert SR, Kalina I, Kang D, Katoh T, Lafuente A, Lin HJ, Romkes M, Taylor JA, Rothman N [2002]. Pooled analysis and meta-analysis of *glutathione S-transferase M1* and bladder cancer: a HuGE review. Am J Epidemiol *156*:95–109; erratum in Am J Epidemiol *156*:492.

Ermolaeva O, Rastogi M, Pruitt KD, Schuler GD, Bittner ML, Chen Y, Simon R, Meltzer P, Trent JM, Boguski MS [1998]. Data management and analysis for gene expression arrays. Nat Genet *20*:19–23.

Ewis AA, Kondo K, Dang F, Nakahori Y, Shinohara Y, Ishikawa M, Baba Y [2006]. *Surfactant protein B* gene variations and susceptibility to lung cancer in chromate workers. Am J Ind Med *49*:367–373.

Faustman EM, Omenn GS [1996]. Risk assessment. In: Klaassen CD, ed. Casarett and Doull's toxicology: the basic science of poisons. 5th ed. New York: McGraw Hill, pp. 75–88.

FDA (Food and Drug Administration) [2007]. Public Health Service Act. Title 42 CFR Part 241. http://www.fda.gov/opacom/laws/phsvcact/sec241.htm. Date accessed October 27, 2008.

FDA (Food and Drug Administration) [1999]. Protection of human subjects. Title 21 CFR Parts 50 and 56. http://www.access.gpo.gov/nara/cfr/waisidx_99/21cfr50_99.html. Date accessed: October 27, 2008.

65 Fed. Reg. 6877 [2000]. The President – executive order 13145: to prohibit discrimination in federal employment based on genetic information. http://www.dol.gov/oasam/regs/statutes/eo13145.htm. Date accessed: February 8, 2008.

74 Federal Register. 9056 [2009]. Regulation under the genetic information nondiscrimination act of 2008. http://edocket.access.gpo.gov/2009/E9-4221.htm Accessed March 2, 2009.

Fong C-S, Wu R-M, Shieh J-C, Chao Y-T, Fu Y-P, Kuao C-L, Cheng C-W [2007]. Pesticide exposure on southwestern Taiwanese with *MnSOD* and *NQO1* polymorphisms is associated with an increased risk of Parkinson's disease. Clin Chem Acta *378*:136–141.

Freeman K [2004]. Toxicogenomics data: the road to acceptance. Environ Health Perspect *112*:A678–A685.

French S [2002]. Genetic testing in the workplace: the employer's coin toss. Duke Law & Technol Rev Sept 15:E1.

Froines JR, Wegman DH, Levenstein C [1988]. The implications of hypersusceptibility for occupational health policy. In: Brain JD, Beck BD, Warren HJ, Shaikh RA, eds. Variations in susceptibility to inhaled pollutants: identification, mechanisms and policy implications. Baltimore, MD: Johns Hopkins Press, pp. 421–442.

Fujisawa T, Ikegami H, Yamato E, Takekawa K, Nakagawa Y, Hamada Y, Oga T, Ueda H, Shintani M, Fukuda M, Ogihara T [1996]. Association of Trp64Arg mutation of the beta3-adrenergic-receptor with NIDDM and body weight gain. Diabetologia *39*:349–352.

GEI (Genetics, Environment and Health Initiative [2007]. The genetics, environment and health initiative: genetics program. http://www.gei.nih.gov/genetics/. Date accessed: March 17, 2008.

Geisler SA, Olshan AF [2001]. *GSTM1*, *GSTT1* and the risk of squamous cell carcinoma of the head and neck: a mini HuGE review. Am J Epidemiol *154*:95–105.

GeneTests [2009]. http://www.genetests.com Date accessed: May 1, 2009.

Genetics and Public Policy Center [2006]. Genetic testing practice guidelines: translating genetic discoveries into clinical care. http://www.dnapolicy.org/resources/Genetic_Testing_Practice_Guideslines.pdf. Date accessed: April 12, 2007.

Gennari L, Merlotti D, De Paola V, Calabro A, Becherini L, Martini G, Nuti R [2005]. Estrogen receptor gene polymorphisms and the genetics of osteoporosis: a HuGE review. Am J Epidemiol *161*:307–320.

Geppert CMA and Roberts LW [2005]. Ethical issues in the use of genetic information in the workplace: a review of recent developments. Curr Opin Psychiatry 18:518–524

Gewirth A [1986]. Human rights and the workplace. Am J Ind Med *9*:31–40.

Gewirth A [1998]. Human rights and genetic testing in the workplace. In: Samuels SW, Upton AC, eds. Genes, cancer and ethics in the work environment. Beverly Farms, MA: OEM Press, pp. 11–24.

Ginsburg GS, Haga SB [2006]. Translating genomic biomarkers into clinically useful diagnostics. Expert Rev Mol Diagn *6*:179–191.

Gledhill BL, Mauro F [1991]. Preface. Gledhill BL, Mauro F, eds. New horizons in biological dosimetry. New York: Wiley-Liss pp. xiv-xx.

Gochfeld M [1998]. Susceptibility biomarkers in the workplace: historical perspective. In: Mendelsohn ML, Mohr LC, Peeters JP, eds. Biomarkers: medical and workplace applications. Washington, DC: Joseph Henry Press, pp. 3–22.

Godderis L, De Boeck M, Haufroid V, Emmery M, Mateuca R, Gardinal S, Kirsch-Volders M, Veulemans H, Lison D [2004]. Influence of genetic polymorphisms on biomarkers of exposure and genotoxic effects in styrene-exposed workers. Environ Molec Mut *44*:293–303.

Godschalk RW, Van Schooten F-J, Bartsch H [2003]. A critical evaluation of DNA adducts as biological markers for human exposure to polycyclic aromatic compounds. J Biochem Mol Biol *36*:1–11.

Goel V [2001]. Appraising organised screening programmes for testing for genetic susceptibility to cancer. BMJ *322*:1174–1178.

Goldstein DB [2009]. Common variation and human traits. N Engl J Med 360:1696–1698.

Goode EL, Ulrich CM, Potter JD [2002]. Polymorphisms in DNA repair genes and associations with cancer risk. Cancer Epidemiol Biomarkers Prev *11*:1513–1530; correction in Cancer Epidemiol Biomarkers Prev *12*:1119.

Grassman JA, Kimmel CA, Neumann DA [1998]. Accounting for variability in responsiveness in human health risk assessment. In: Neumann DA, Kimmel CA, eds. Human variabililty in response to chemical exposures: measures, modeling and risk assessment. Boca Raton, FL: CRC Press. pp. 1-26.

Greenland S [1993]. Basic problems in interaction assessment. Environ Health Perspect *101*(S4):59–66.

Grodsky JA [2007]. Genomics and toxic torts: dismantling the risk-injury divide. Stanford Law Rev *59*:1671-1734.

Grody WW [2003]. Ethical issues raised by genetic testing with oligonucleotide microarrays. Mol Biotechnol *23*:127–138.

Groopman JD, Kensler TW [1999]. The light at the end of the tunnel for chemical-specific biomarkers: daylight or head light? Carcinogenesis *20*:1–11.

Gwosdz C, Balz V, Scheckenbach K, Bier H [2005]. p53, p63 and p73 expression in squamous cell carcinomas of the head and neck and their response to cisplatin exposure. Adv Otorhinolaryngol *62*:58–71.

Hagmar L, Bonassi S, Strömberg U, Brøgger A, Knudsen L, Norppa H, Reuterwall C [1998]. Chromosomal aberrations in lymphocytes predict human cancer: a report from the European study group on cytogenetic biomarkers and health (ESCH). Cancer Res *58*:4117–4121.

Hagmar L, Brogger A, Hansteen IL, Heim S, Hogstedt B, Knudsen L, Lambert B, Linnainmaa K, Mitelman F, Nordenson I et al. [1994]. Cancer risk in human predicted by increased levels of chromosomal aberrations in lymphocytes: Nordic study group on health risks of chromosome damage. Cancer Res *54*:2919–2922.

Hagmar L, Stromberg U, Bonassi S, Hansteen IL, Knudsen LE, Lindholm C, Norppa H [2004]. Impact of types of lymphocyte chromosomal aberrations on human cancer risk: results from Nordic and Italian cohorts. Cancer Res *64*:2258–2263.

Hagmar L, Wirfalt E, Paulsson B, Tornqvist M [2005]. Differences in hemoglobin adduct levels of acrylamide in the general population with respect to dietary intake, smoking habits and gender. Mutat Res *580*:157–165.

Haldane JBS [1938]. The biology of inequality. In: Heredity and politics. London: Allen and Unwin, pp. 179–180.

Halperin WE, Frazier TM [1985]. Surveillance for the effects of workplace exposure. Ann Rev Public Health *6*:419–432.

Hanash S [2003]. Disease proteomics. Nature *422*:226–232.

Hattis D [1998]. Strategies for assessing human variability in susceptibility and using variability to infer human risks. In: Newmann DA, Kimmel CA, eds. Human variability in response to chemical exposures: measures, modeling and risk assessment. Boca Raton, FL: CRC Press, pp. 27–57.

Hattis D, Silver K [1993]. Use of biomarkers in risk assessment: practical applications. In: Schulte PA, Perera FR, eds. Molecular epidemiology: principles and practices. San Diego, CA: Academic Press, pp. 251–273.

Hattis D, Swedis S [2001]. Uses of biomarkers for genetic susceptibility and exposure in the regulatory context. Jurimetrics *41*:177–194.

Henry CJ, Phillips R, Carpanini F, Corton JC, Craig K, Igarashi K, Leboeuf R, Marchant G, Osborn K, Pennie WD, Smith LL, Teta MJ, Vu V [2002]. Use of genomics in toxicology and epidemiology: findings and recommendations of a workshop. Environ Health Perspect *110*:1047–1050.

HGP (Human Genome Project) [2006]. National Human Genome Research Institute, National Institutes of Health. Genetic testing report, chapter 2. http://www.genome.gov/10002404. Date accessed: October 24, 2006.

HGP (Human Genome Project) [2009]. Insights learned from the human DNA sequence. http://www.ornl.gov/sci/techresources/Human-Genome/project/journals/insight.shtml Date accessed: April 23, 2009.

Hinney A, Nguyen TT, Scherag A, Friedel S, Brönner G, Müller TD, Grallert H, Illig T, Wichmann HE, Rief W, Schäfer H, Hebebrand J [2007]. Genome wide association (gwa) study for early onset extreme obesity supports the role of fat mass and obesity associated gene *(FTO)* variants. PLoS ONE 2:e1361.

HIPAA (Health Insurance Portability and Accountability Act) [2005]. http://www.pinsco.com/downloads/HIPAA_Privacy_Rule_Alert_Mar_05.pdf. Date accessed: June 28, 2006.

HIPAA (Health Insurance Portability and Accountability Act) [1996]. http://cms/hhs/gov/HIPAAGenInfo/Downloads/HIPAALaw.pdf/. Date accessed: June 26, 2006.

Hirvonen A, Pelin K, Tammilehto L, Karjalainen A, Mattson K and Linnainmaa K [1995]. Inherited *GSTM1* and *NAT2* defects as concurrent risk modifiers in asbestos-related human malignant mesothelioma. Cancer Res *55*:2981–2983.

Hirschhorn JN [2009]. Genome-wide association studies- illuminating biologic pathways. N Engl J Med *360*:1699–1701.

Hodge JG Jr. [1998]. Privacy and antidiscrimination issues: genetics legislation in the United States. Community Genet *1*:169–174.

Holeckova B, Piesova E, Sivikova K, Dianovsky J [2004]. Chromosomal aberrations in humans induced by benzene. Ann Agric Environ Med *11*:175–179.

Holland NT, Smith MT, Eskenazi B, Bastaki M [2003]. Biological sample collection and processing for molecular epidemiological studies. Mutat Res *543*:217–234.

Holtzman NA [2003]. Ethical aspects of genetic testing in the workplace. Community Genet *6*:136–138.

Holtzman NA, Marteau TM [2000]. Will genetics revolutionize medicine? N Engl J Med *343*:141–144.

Hornig DF [1988]. Conclusion. In: Brain JD, Beck BD, Warren AJ, Shaikh RA, eds. Variations in susceptibility to inhaled pollutants: identification, mechanisms and policy implications. Baltimore, MD: Johns Hopkins University Press, pp. 461–471.

Hsieh H-I, Chen P-C, Wong R-H, Wang J-D, Yang P-M, Cheng T-J [2007]. Effect of the *CYP2E1* genotype on vinyl chloride monomer-induced liver fibrosis among polyvinyl chloride workers. Toxicology *239*:34-44.

HuGENet [2009]. http://www.cdc.gov/genomics/hugenet/about.htm Date accessed: March 19, 2009.

Hung RJ, Boffetta P, Brennan P, Malaveille C, Gelatti U, Placidi D, Carta A, Hautefeuille A, Porru S [2004a]. Genetic polymorphisms of *MPO*, *COMT*, *MnSOD*, *NQO1*, interactions with environmental exposures and bladder cancer risk. Carcinogenesis *25*:973–978.

Hung RJ, Boffetta P, Brennan P, Malaveille C, Hautefeuille A, Donato F, Gelatti U, Spaliviero M, Placidi D, Carta A, Scotto di Carlo A, Porru S [2004b]. *GST*, *NAT*, *SULT1A1*, *CYP1B1* genetic polymorphisms, interactions with environmental exposures and bladder cancer risk in a high-risk population. Int J Cancer *110*:598–604.

Hung RJ, Boffetta P, Canzian F, Moullan N, Szeszenia-Dabrowska N, Zaridze D, Lissowska J, Rudnai P, Fabianova E, Mates D, Foretova L, Janout V, Bencko V, Chabrier A, Landi S, Gemignani F, Hall J, Brennan P [2006]. Sequence variants in cell cycle control pathway, x-ray exposure, and lung cancer risk: a multicenter case-control study in central Europe. Cancer Res *66*:8280–8286.

Hunter DJ [2005]. Gene-environment interactions in human diseases. Nat Rev Genet *6*:287–298.

Hunter D, Caporaso N [1997]. Informed consent in epidemiologic studies involving genetic markers. Epidemiology *8*:596–599.

Hunter DJ, Khoury MJ, Drazen JM [2008]. Letting the genome out of the bottle – will we get our wish? N Engl J Med *358*:105-107.

Hunter DJ, Thomas G, Hoover RN, Chanock SJ [2007]. Scanning the horizon: what is the future of genome-wide association studies in accelerating discoveries in cancer etiology and prevention? Cancer Causes and Control *18*:479-484.

Hustead JL, Goldman J [2002]. Genetics and privacy. Am J Law Med *28*:285–307.

IARC [2004]. IARC mechanisms of carcinogenesis. In: Vineis P, Schulte PA, Carreón T, Bailer AJ, Buffler P, Rice J, Bird M, Boffetta P, eds. IARC Scientific Publication No. 157. Lyon, France: International Agency for Research on Cancer, pp. 417–435.

IARC [1997]. IARC monographs on the evaluation of carcinogenic risks to humans: some industrial chemicals. Vol. 142. Lyon, France: World Health Organization, International Agency for Research on Cancer Scientific Publications.

IARC [1994]. IARC monographs on the evaluation of carcinogenic risks to humans. some industrial chemicals. Vol. 60. Lyon, France: World Health Organization, International Agency for Research on Cancer Scientific Publications, pp. 73–159.

IHC (International HapMap Consortium) [2007]. A second generation human haplotype map of over 3.1 million SNPs. Nature *449*:851-861.

IHC (International HapMap Consortium) [2005]. A haplotype map of the human genome. Nature *437*:1299–1320.

IHC (International HapMap Consortium) [2003]. The International HapMap Project. Nature *426*:789-96.

Ioannidis JP, Bernstein J, Boffetta P, Danesh J, Dolan S, Hartge P, Hunter D, Inskip P, Jarvelin M-R, Little J, Maraganore DM, Newton-Bishop JA, O'Brien TR, Petersen G, Riboli E, Seminara D, Taioli E, Uitterlinden AG, Vineis P, Winn DM, Salanti G, Higgins JPT, Khoury, MJ [2005]. A network of investigator networks in human genome epidemiology. Am J Epidemiol *162*:302–304.

Ioannidis JP, Gwinn M, Little J, Higgins JP, Bernstein JL, Boffetta P, Bondy M, Bray MS, Benchley PE, Buffler PA, Casas JP, Chokkalingam A, Danesh J, Davey Smith G, Dolan S, Duncan R, Gruis NA, Hartge P, Hashibe M, Hunter DJ, Jarvelin M-R, Malmer B, Maraganore DM, Newton-Bishop JA, O'Brien TR, Petersen G, Riboli E, Salanti G, Seminara D, Smeeth L, Taioli E, Timpson N, Uitterlinden AG, Vineis P, Wareham N, Winn DM, Zimmern R, Khoury, MJ, Human Genome Epidemiology Network and the Network of Investigator Networks [2006]. A roadmap for efficient and reliable human genome epidemiology. Nature Gen *38*:3–5.

IPCS (International Programme on Chemical Safety) [2001]. Biomarkers in risk assessment: validity and validation. Environmental Health Criteria #222. http://www.inchem.org/documents/ehc/ehc/ehc222.htm Date accessed: October 26, 2006.

Javitt GH [2006]. Policy implications of genetic testing: not just for geneticists anymore. Adv Chronic Kidney Disease *13*:178–182.

Javitt GH, Hudson K [2006]. Federal neglect: regulation of genetic testing. Issues Sci Technol *22*:59–66.

Jones IM, Galick H, Kato P, Langlois RG, Mendelsohn ML, Murphy GA, Pleshanov P, Ramsey MJ, Thomas CB, Tucker JD, Tureva L, Vorobtsova I, Nelson DO [2002]. Three somatic genetic biomarkers and covariates in radiation-exposed cleanup workers of the Chernobyl nuclear reactor 6–13 years after exposure. Radiat Res *158*:424–442.

Jones IM, Tucker JD, Langlois RG, Mendelsohn ML, Pleshanov P, Nelson DO [2001]. Evaluation of three somatic genetic biomarkers as indicators of low dose radiation effects in clean-up workers of the Chernobyl nuclear reactor accident. Radiat Prot Dosimetry *97*:61–67.

Joo W-A, Sul D, Lee D-Y, Lee E, Kim C-W [2004]. Proteomic analysis of plasma proteins of workers exposed to benzene. Mutat Res *558*:35–44.

Jorde LB, Carey JC, White RL [1997]. Medical genetics. St. Louis, MO: Mosby-Year Book, Inc., pp. 127, 268.

Kalsheker N, Morgan K [1990]. Molecular biology and respiratory disease. the α_1–*antitrypsin* gene and chronic lung disease. Thorax *45*:759–764.

Karlin S, Taylor HM [1975]. A first course in stochastic processes. Boston, MA: Academic Press.

Keavney B, McKenzie C, Parish S, Palmer A, Clark S, Youngman L, Delepine M, Lathrop M, Peto R, Collins R [2000]. Large-scale test of hypothesized associations between the angiotensin-converting-enzyme insertion/deletion polymorphism and myocardial infarction in about 5000 cases and 6000 controls. Lancet *355*:434–442.

Kelada SN, Eaton DL, Wang SS, Rothman NR, Khoury MJ [2003]. The role of genetic polymorphisms in environmental health. Env Health Perspect *111*:1055–1064.

Kelada SN, Shelton E, Kaufmann RB, Khoury MJ [2001]. δ-*Aminolevulinic acid dehydratase (ALAD)* genotype and lead toxicity: a HuGE review. Am J Epidemiol *154*:1–13.

Kellen E, Hemelt M, Broberg K, Golka K, Kristensen VN, Hung RJ, Matullo G, Mittal RD, Porru S, Povey A, Schulz WA, Shen J, Buntinx F, Zeegers MP, Taioli E [2007]. Pooled analysis and meta-analysis of the *glutathione S-transferase P1 Ile 105Val* polymorphism and bladder cancer: a HuGE-GSEC review. Am J Epidemiol *165*:1221–1230.

Kennedy S [2002]. The role of proteomics in toxicology: identification of biomarkers of toxicity by protein expression analysis. Biomarkers *7*:269–290.

Kenneson A, Van Naarden-Braun K, Boyle C [2002]. *GJB2 (connexin 26)* variants and nonsyndromic sensorineural hearing loss: a HuGE review. Genet Med *4*:258–274.

Keohavong P, Lan Q, Gao W-M, Zheng K-C, Mady HH, Melhem MF, Mumford JL [2005]. Detection of *p53* and *K-ras* mutations in sputum of individuals exposed to smoky coal emissions in Xuan Wei County, China. Carcinogenesis *26*:303–308.

Keshava C, McCanlies EC, Weston A [2004]. *CYP3A4* polymorphisms: potential risk factors for breast and prostate cancer: a HuGE review. Am J Epidemiol *160*:825–841.

Khoury MJ [2002]. Commentary: Epidemiology and the continuum from genetic research to genetic testing. Am J Epidemiol *156:* 297–299.

Khoury MJ, Beaty TH, Cohen BH [1993]. Applications of genetic epidemiology in medicine and public health. In: Fundamentals of genetic epidemiology (monographs in epidemiology and biostatistics). Vol. 22. New York, NY: Oxford University Press, pp. 312–339.

Khoury MJ, Little J [2000]. Human genome epidemiologic reviews: the beginning of something HuGE. Am J Epidemiol *151*:2–3.

Khoury MJ, Newill CA, Chase GA [1985]. Epidemiologic evaluation of screening for risk factors: application to genetic screening. Am J Public Health *75*:1204–1208.

Khoury MJ, Wagener DK [1995]. Epidemiological evaluation of the use of genetic to improve the predictive value of disease risk factors. Am J Hum Genet *56*:835–844.

Kidd JM, Cooper GM, Donahue WF, Hayden HS, Sampas N, Graves T, Hansen N, Teague B, Alkan C, Antonacci F, Haugen E, Zerr T, Yamada NA, Tsang P, Newman TL, Tu"zu"n E, Cheng A, Ebling HM, Tusneem N, David R, Gillett W, Phelps KA, Weaver M, Saranga D, Brand A, Tao W, Gustafson E, McKernan K, Chen L, Malig M, Korn JM, McCarroll SA, Altshuler DA, Peiffer DA, Dorschner M, Stamatoyannopoulos J, Schwartz D, Nickerson DA, Mullikin JC, Wilson RK, Bruhn L, Olson MV, Kaul R, Smith DR, Eichler EE [2008]. Mapping and sequencing of structural variation form eight human genomes. Nature *453*:56–64.

Kilian DJ, Picciano DJ [1979]. Monitoring for chromosomal damage in exposed industrial populations. In: Berg K, ed. Genetic damage in man caused by environmental agents. New York: Academic Press, pp. 101–115.

Kim P [2002]. Genetic discrimination, genetic privacy: rethinking employee protections for a brave new workplace. Faculty working paper series, paper No. 02–10–02, October 2002. St. Louis, MO: Washington University School of Law.

King HC, Sinha AA [2001]. Gene expression profile analysis by DNA microarrays: promise and pitfalls. JAMA *286*:2280–2288.

Kline JN, Doekes G, Bonlokke J, Hoffman HJ, Von Essen S, Zhai R [2004]. Working group report 3: sensitivity to organic dusts – atopy and gene polymorphisms. Am J Ind Med *46*:416–418.

Koizumi S [2004]. Application of DNA microarrays in occupational health research. J Occup Health *46*:20–25.

Kraft P, Hunter DJ [2009]. Genetic risk prediction—are we there yet? N Engl J Med 360:1701–1703.

Kreiss K, Mroz MM, Newman LS, Martyny J, Zhen B [1996]. Machining risk of beryllium disease and sensitization with median exposures below 2 micrograms/m^3. Am J Ind Med *30*:16–25.

Krumm J [2002]. Genetic discrimination: why Congress must ban genetic testing in the workplace. J Leg Med *23*:491–521.

Kulynych J, Korn D [2002]. Use and disclosure of health information in genetic research: weighing the impact of the new federal medical privacy rule. Am J Law Med *28*:309–324.

Langer CS [1996]. Title 1 of the Americans With Disabilities Act. Occup Med *11*:5–16.

Langlois RG, Bigbee WL, Kyoizumi S, Nakamura N, Bean MA, Akiyama M, Jensen RH [1987]. Evidence for increased somatic cell mutations at the glycophorin A locus in atomic bomb survivors. Science *236*:445–448.

Lappe M [1983]. Ethical issues in testing for differential sensitivity to occupational hazards. J Occup Med *25*:797–808.

Last JM, ed. [1988]. A dictionary of epidemiology. 2nd ed. New York: Oxford University Press.

Launis V [2000]. The use of genetic test information in insurance: the argument from indistinguishability reconsidered. Sci Eng Ethics *6*:299–310.

Lee JW, Weiner RS, Sailstad JM, Bowsher RR, Knuth DW, O'Brien PJ, Fourcroy JL, Dixit R, Pandite L, Pietrusko RG, Soares HD, Quarmby V, Vesterqvist OL, Potter DM, Witliff JL, Fritche HA, O'Leary T, Perlee L, Kadam S, Wagner JA [2005]. Method validation and measurement of biomarkers in nonclinical and clinical samples in drug development: a conference report. Pharm Res *22*:499–511.

Legator MS [1995]. Application of biomarkers: getting our priorities straight. In: Mendelsohn ML, Peeters JP, Normandy MJ, eds. Biomarkers and occupational health: progress and perspectives. Washington, DC: Joseph Henry Press, pp. 61–69.

Lemmens T [1997]. What about your genes? Ethical, legal and policy dimensions of genetics in the workplace. Pol Life Sci *16*:57–75.

Lemmens T, Freedman B [2000]. Ethics review for sale? Conflict of interest and commercial research review boards. Millbank Quarterly *78*:547-584.

Lin G, Guo W, Chen J, Qin Y, Golka K, Xiang C, Ma Q, Lu D, Shen J [2005]. An association of *UDP-glucuronosyltransferase 2B7 C802T (His268Tyr)* polymorphism with bladder cancer in benzidine-exposed workers in China. Toxicol Sci *85*:502–506.

Liou SH, Lung JC, Chen YH, Yang T, Hsieh LL, Chen CJ, Wu TN [1999]. Increased chromosome-type chromosome aberration frequencies as biomarkers of cancer risk in a Blackfoot endemic area. Cancer Res *59*:1481–1484.

Little J [2004]. Reporting and review of human genome epidemiology studies. In: Khoury ML, Little J, Burke W, eds. Human genome epidemiology: a scientific foundation for using genetic information to improve health and prevent disease. New York: Oxford University Press, pp. 168–192.

Little J, Khoury MJ [2003]. Mendelian randomization: a new spin or real progress? Lancet *362*:930–931.

Little J, Bradley L, Bray MS, Clyne M, Dorman J, Ellsworth DL, Hanson J, Khoury M, Lau J, O'Brien TR, Rothman N, Stroup D, Taioli E, Thomas D, Vainio H, Wacholder S, Weinberg C [2002]. Reporting, appraising, and integrating data on genotype prevalence and gene-disease associations. Am J Epidemiol *156*:300–310.

Lu J, Jin T, Nordberg G, Nordberg M [2001]. Metallothionein gene expression in peripheral lymphocytes from cadmium-exposed workers. Cell Stress Chaperones *6*:97–104.

Ma QW, Lin GF, Chen JG, Xiang CQ, Guo WC, Golka K, Shen JH [2004]. Polymorphism of *N-acetyltransferase 2 (NAT2)* gene polymorphism in shanghai population: occupational and non-occupational bladder cancer patient groups. Biomed Environ Sci *17*(3):291–298.

MacDonald C, Williams-Jones B [2002]. Ethics and genetics: susceptibility testing in the workplace. J Bus Ethics *35*:235–241.

Malaspina A [1998]. Foreword. In: Neumann D, Kimmel CA, eds. Human variability in response to chemical exposures: measures, modeling and risk assessment. Boca Raton, FL: CRC Press.

Maltby L [2000]. Brave new workplace: genetic breakthroughs and the new employment discrimination. Lynda M. Fox memorial keynote address presented at the Given Institute, University of Colorado School of Medicine, Aspen, CO, July 21, 2000. http://www.workrights.org/issue_genetic/gd_brave_new_speech.html. Date accessed: August 26, 2007.

Mapp CE, Beghe B, Balboni A, Zamorani G, Padoan M, Jovine L, Baricordi OR, Fabbri LM [2000]. Association between *HLA* genes and susceptibility to toluene diisocyanate-induced asthma. Clin Exp Allergy *30*:651–656.

Mapp CE, Fryer AA, DeMarzo N, Pozzato V, Padoan M, Boschetto P, Strange RC, Hemmingsen A, Spiteri MA [2002]. *Glutathione S-transferase GSTP1* is a susceptibility gene for occupational asthma induced by isocyanates. J Allergy Clin Immunol *109*:867–872.

Marchant GE [2000]. Genetic susceptibility and biomarkers in toxic injury litigation. Jurimetrics *41*:67–109.

Marchant GE [2003a]. Genomics and toxic substances. Part 1: Toxicogenomics. Environ Law Rep *33*:10071–10093.

Marchant GE [2003b]. Genomics and toxic substances. Part 2: Genetic susceptibility to environmental agents. Environ Law Rep *33*:10641–10667.

Marcus PM, Hayes RB, Vineis P, Garcia-Closas M, Caporaso NE, Autrup H, Branch RA, Brockmoller J, Ishizaki T, Karakaya AE, Ladero JM, Mommsen S, Okkels H, Romkes M,

Roots I, Rothman N [2000a]. Cigarette smoking, N-acetyltransferase 2 acetylation status, and bladder cancer risk: a case-series meta-analysis of a gene-environment interaction. Cancer Epidemiol Biomarkers Prev 9:461–467.

Marcus PM, Vineis P, Rothman N [2000b]. NAT2 slow acetylation and bladder cancer risk: a meta-analysis of 22 case-control studies conducted in the general population. Pharmacogenetics 10:115–122.

Marczynski B, Raulf-Heimsoth M, Preuss R, Kappler M, Schott K, Pesch B, Zoubek G, Hahn JU, Mensing T, Angerer J, Kafferlein HU, Bruning T [2006]. Assessment of DNA damage in WBCs of workers occupationally exposed to fumes and aerosols of bitumen. Cancer Epidemiol Biomarkers Prev 15:645-651.

Marteau T, Richards M [1996]. The troubled helix: social and psychological implications of the new human genetics. Cambridge, MA: University Press.

McCanlies EC, Kreiss K, Andrew M, Weston A [2003]. *HLA-DPB1* and chronic beryllium disease: a HuGE review. Am J Epidemiol 157:388–398.

McCanlies EC, Landsittel DP, Yucesoy B, Vallyathan V, Luster ML, Sharp DS [2002]. Significance of genetic information in risk assessment and individual classification using silicosis as a case model. Ann Occup Hyg 46:375–381.

McCanlies EC, Schuler CR, Kreiss K, Frye BL, Ensey JS, Weston A [2007]. TNF-alpha polymorphisms in chronic beryllium disease and beryllium sensitization. J Occup Environ Med 49:446-52.

McCanlies E, Weston A [2004]. Immunogenetics of chronic beryllium disease. In: Khoury ML, Little J, Burke W, eds. Human genome epidemiology: a scientific foundation for using genetic information to improve health and prevent disease. New York: Oxford University Press, pp. 383–401.

McCunney RJ [2002]. Genetic testing: ethical implications in the workplace. Occup Med 17:665–672.

Medeiros MG, Rodrigues AS, Batoreu MC, Laires A, Rueff J, Zhitkovich A [2003]. Elevated levels of DNA-protein crosslinks and micronuclei in peripheral lymphocytes of tannery workers exposed to trivalent chromium. Mutagenesis 18:19–24.

Mendelsohn ML [1995]. The current applicability of large-scale biomarker programs to monitor cleanup workers. In: Mendelsohn ML, Peeters JP, Normandy MJ, eds. Biomarkers and occupational health: progress and perspectives. Washington, DC: Joseph Henry Press, pp. 9–19.

Migliore L, Naccarati A, Coppede F, Bergamaschi E, De Palma G, Voho A, Manini P, Jarventaus H, Mutti A, Norppa H, Hirvonen A [2006]. Cytogenetic biomarkers, urinary metabolites and metabolic gene polymporphisms in workers exposed to styrene. Pharmacogenet Genomics *16*:87-99.

Milacic S [2005]. Frequency of chromosomal lesions and damaged lymphocytes of workers occupationally exposed to x rays. Health Phys *88*:334–339.

Millikan R [2002]. The changing face of epidemiology in the genomics era. Epidemiology *13*:472-480.

Mitchell CS [2002]. Confidentiality in occupational medicine. Occup Med *17*:617–623.

Mitchell RJ, Farrington SM, Dunlop MG, Campbell H [2002]. Mismatch repair genes *hMLH1* and *hMSH2* and colorectal cancer: a HuGE review. Am J Epidemiol *156*:885–902.

Modugno F [2004]. Ovarian cancer and polymorphisms in the androgen and progesterone receptor genes: a HuGE review. Am J Epidemiol *159*:319–335.

Moore DH II, Tucker JD [1999]. Biological dosimetry of Chernobyl cleanup workers: inclusion of data on age and smoking provides improved radiation dose estimates. Radiat Res *152*:655–664.

Morgan KT, Brown HR, Benavides G, Crosby L, Sprenger D, Yoon L, Ni H, Easton M, Morgan D, Laskowitz D, Tyler R [2002]. Toxicogenomics and human disease risk assessment. Hum Ecol Risk Assess *8*:1339–1353.

Morgenstern H, Thomas D [1993]. Principles of study design in environmental epidemiology. Environ Health Perspect *101*:23–28.

Mrdjanovic J, Jakimov D, Tursijan S, Bogdanovic G [2005]. Evaluation of sister chromatid exchanges, micronuclei, and proliferating rate index in hospital workers chronically exposed to ionizing radiation. J BOUN *10*:99-103.

Murray TH [1983]. Genetic screening in the workplace: ethical issues. J Occup Med *25*:451–454.

Murray TH [1997]. Genetic exceptionalism and future diaries: is genetic information different from other medical information. In: Rothstein MA, ed. Genetic secrets: protecting privacy and confidentiality in the genetic era. New Haven, CT: Yale University Press, pp. 60–73.

Nahed BV, Bydon M, Ozturk AK, Bilguvar K, Bayrakli F, Gunel M [2007]. Genetics of intracranial aneurysms. Neurosurgery *60*: 213-225.

Nakayama EE, Meyer L, Iwamoto A, Persoz A, Nagai Y, Rouzioux C, Delfraissy JF, Debre P, McIlroy D, Theodorou I, Shioda T; SEROCO Study Group [2002]. Protective effect of *interleukin-4 -589T* polymorphism on human immunodeficiency virus type 1 disease progression: relationship with virus load. J Infect Dis *185*:1183–1186.

NAS (National Academy of Sciences) [2003]. Focused efforts in toxigenomics. Committee on emerging issues and data on environmental contaminants. Issue 3, July 2. http://dels.nas.edu/emergingissues/docs/EI_issue3.pdf. Date accessed: August 11, 2006.

National Conference of State Legislatures [2004]. http://www.ncsl.org/public/leglinks.cfm. Date accessed July 14, 2009.

NBAC (National Bioethics Advisory Commission) [1999]. http://bioethics.georgetown.edu/nbac/pubs.html. Date accessed: August 12, 2007.

NCD (National Council on Disability) [2003]. No.9 Chevron v. Echazabal: The ADA's "direct threat to self" defense. http://www.ncd.gov/newsroom/publications/2003/directthreat.htm. Date accessed: February 20, 2009.

NCI (National Cancer Institute) [2006]. Cancer Genetics Overview (PDQ^R) Health Professional Version. http://www.cancer.gov/cancertopics/pdq/genetics/overview/healthprofessional. Date accessed: October 26, 2006.

Nebert DW, Roe AL, Vandale SE, Bingham E, Oakley GG [2002]. *NAD(P)H: quinone oxidoreductase (NQO1)* polymorphism, exposure to benzene, and predisposition to disease: a HuGE review. Genet Med *4*:62–70.

Nederhand RJ, Droog S, Kluft C, Simoons ML, de Maat MP; investigators of the EUROPA trial [2003]. Logistics and quality control for DNA sampling in large multicenter studies. J Thromb Haemost *1*:987–991.

Nelson KA, Witte JS [2002]. Androgen receptor CAG repeats and prostate cancer. Am J Epidemiol *155*:883–890.

Neumann DA, Kimmel CA, eds. [1998]. Human variability in response to chemical exposures: measures, modeling and risk assessment. Washington, DC: International Life Sciences Institute/CRC Press.

NHGRI (National Human Genome Research Institute) [2004]. http://www.nhgri.nih.gov. Date accessed: March 12, 2005.

NIH (National Institutes of Health) [2007]. Certificates of confidentiality. http://grants1.nih.gov/grants/policy/coc/background.htm. Date accessed: November 2, 2008.

NIH (National Institutes of Health) [2004]. Protecting personal health information in research: understanding the HIPAA Privacy Rule. Rev. Rockville, MD: U.S. Department of Health and Human Services, Centers for Disease Control and Prevention, National Institutes of Health, Publication No. 03–5–388. http://privacyruleandresearch.nih.gov/pr_02.asp. Date accessed: November 2, 2008.

NRC (National Research Council) [2007]. Applications of toxigenomic technologies to predictive toxicology and risk assessment. National Academies Press, Washington, D.C.

NRC (National Research Council) [1987]. Biological markers in environmental health research. Env Health Perspect *74*:3–9.

Ntais C, Polycarpou A, Ioannidis JPA [2004]. Meta-analysis of the association of the cathepsin D Ala224Val gene polymorphism with the risk of Alzheimer's disease: a HuGE gene-disease association review. Am J Epidemiol *159*:527–536.

Nuremberg Code [1949]. Trials of war criminals before the Nuremberg military tribunals under control council law No. 10, Vol. 2. Washington, DC: U.S. Government Printing Office, pp. 181–182. http://ohsr.od.nih.gov/guidelines/nuremberg.html. Date accessed: November 2, 2008.

Nuwaysir EF, Bittner M, Trent J, Barrett JC, Afshari CA [1999]. Microarrays and toxicology: the advent of toxigenomics. Mol Carcinog *24*:153–159.

Oizumi T, Daimon M, Saitoh T, Kameda W, Yamaguchi H, Ohnuma H, Igarashi M, Eguchi H, Manaka H, Tominaga M, Kato T [2001]. Genotype Arg/Arg, but not Trp/Arg, of the Trp64Arg polymorphism of the β3-adrenergic receptor is associated with Type 2 diabetes and obesity in a large Japanese sample. Diabetes Care *24*:1579–1583.

Olshan AF, Weissler MC, Watson MA, Bell DA [2000]. *GSTM1, GSTT1, GSTP1, CYP1A1,* and *NAT1* polymorphisms, tobacco use, and the risk of head and neck cancer. Cancer Epidemiol Biomarkers Prev *9*:185–191.

Omenn GS [1982]. Predictive identification of hypersusceptible individuals. J Occup Med *24*:369–374.

Omori K, Kazama JJ, Song J, Goto S, Takada T, Saito N, Sakatsume M, Narita I, Geyjo F [2002]. Association of the MCP-1 gene polymorphism A-2518G with carpal-tunnel syndrome in hemodialysis patients. Amyloid *9*:175–182.

OSHA (Occupational Safety and Health Administration) [2007a] Cadmium - 1910.1027. http://www.osha.gov/SLTC/cadmium/standards.html. Date accessed: February 20, 2009.

OSHA (Occupational Safety and Health Administration) [2007b]. http://www.osha.gov/SLTC/lead/standards.html. Date accessed: February 20, 2009.

OSHA (Occupational Safety and Health Administration) [1980a]. Press release USDL 80–107. February 20.

OSHA (Occupational Safety and Health Administration) [1980b]. OSHA clarifies references to genetic factors in medical surveillance requirements. News, U.S. Department of Labor, 80–107, February 20.

OSHA (Occupational Safety and Health Administration) [1974]. 29 CFR Part 1900 to end. Washington DC: U.S. Government Printing Office, pp. 131–201.

OTA (Office of Technology Assessment) [1990]. Genetic monitoring and screening in the workplace. Washington, DC: Congress of the United States, OTA–B4–455. http://www.access.gpo.gov/ota/. Date accessed: April 13, 2006.

OTA (Office of Technology Assessment) [1983]. The role of genetic testing on the prevention of occupational disease. Washington, DC: Congress of the United States, OTA–BA–194. http://www.access.gpo.gov/ota/. Date accessed: April 27, 2006.

Ottman R [1996]. Gene-environment interaction: definitions and study designs. Prev Med 25:764–770.

Paracchini V, Raimondi S, Gram IT, Kang D, Kocabas NA, Kristensen VN, Li D, Parl FF, Rylander-Rudqvist T, Soucek P, Zheng W, Wedren S, Taioli E [2007]. Meta- and pooled analyses of the *cytochrome P-450 1B1 Val432Leu* polymorphism and breast cancer: a HuGE-GSEC review. Am J Epidemiol 165:115–125.

Parker L, Majeske RA [1996]. Standards of care and ethical concerns in genetic testing and screening. Clin Obstet Gynecol 39:873–884.

Pavanello S, Clonfero E [2004]. Individual susceptibility to occupational carcinogens: the evidence from biomonitoring and molecular epidemiology studies. G Ital Med Lav Ergon 26:311–321.

Pavanello S, Kapka L, Siwinska E, Mieizynska D, Bolognesi C, Clonfero E [2008]. Micronuclei related to anti-B[a]PDE-DNA adduct in peripheral blood lymphocytes of heavily polycyclic aromatic hydrocarbon-exposed nonsmoking coke-oven workers and controls. Cancer Epidemiol Biomarkers Prev 17:2795-2799.

Peltonen L, McKusick VA [2001]. Genomics and medicine: dissecting human disease in the postgenomic era. Science 291:1224–1229.

Perera FP, Hemminki K, Gryzbowska E, Motykiewicz G, Michalska J, Santella RM, Young TL, Dickey C, Brandt-Rauf P, DeVivo I, Blaner W, Tsai WY, Chorazy M [1992]. Molecular and genetic damage in humans from environmental pollution in Poland. Nature 360: 256–258; erratum in Nature 361:564.

Perera FP, Mooney LA, Stampfer M, Phillips DH, Bell DA, Rundle A, Cho S, Tsai WY, Ma J, Blackwood A, Tang D, [2003]. Associations between carcinogen-DNA damage, glutathione S-transferase genotypes, and risk of lung cancer in the prospective Physicians' Health Cohort Study. Carcinogenesis 23:1641–1646.

Perera FP, Weinstein IB [1982]. Molecular epidemiology and carcinogen-DNA adduct detection: new approaches to studies of human cancer causation. J Chronic Dis 35:581–600.

Peters S, Talaska G, Jonsson BA, Kromhout H, Vermeulen R [2008]. Polycyclic aromatic hydrocarbon exposure, urinary mutagenicity, and DNA adducts in rubber manufacturing workers. Cancer Epidemiol Biomarkers Prev 17:1452-1459.

Piirila P, Wikman H, Luukkonen R, Kaaria K, Rosenberg C, Nordman H, Norppa H, Vainio H, Hirvonen A [2001]. *Glutathione S-transferase* genotypes and allergic responses to diisocyanate exposure. Pharmacogenetics 11:437–445.

Plog BA, Quinlan PJ, eds. [2002]. Fundamentals of industrial hygiene, 5th ed. Itasca, NY: National Safety Council Press pp.3-32.

Poirier MC, Weston A [2002]. DNA damage, DNA repair, and mutagenesis. In: Bertino J, ed. Encyclopedia of cancer. 2nd ed. Vol. 1. San Diego, CA: Academic Press, pp. 641–649.

Ponce RA, Bartell SM, Kavanagh TK, Woods JS, Griffith WC, Lee RC, Takaro TK, Faustman EM [1998]. Uncertainty analysis for comparing predictive models and biomarkers: a case study of dietary methyl mercury exposure. Regul Toxicol Pharmacol 28:96–105.

Poulter SR [2001]. Genetic testing in toxic injury litigation: the path to scientific certainty or blind alley? Jurimetrics J 41:211–238.

Press N, Burke W [2001]. Genetic exceptionalism and the paradigm of risk in United States biomedicine. A Decade of ELSI Research: A Celebration of the First Ten Years of the Ethical, Legal, and Social Implications Program (Bethesda, MD, January 16–18, 2001).

Purcell S, Cherny SS, Sham PC [2003]. Genetic power calculator: design of linkage and association genetic mapping studies of complex traits. Bioinformatics 19:149–150.

Pylkkanen L, Sainio M, Ollikainen T, Mattson K, Nordling S, Carpen O, Linnainmaa K, Husgafvel-Pursiainen K [2002]. Concurrent LOH at multiple loci in human malignant mesothelioma with preferential loss of *NF2* gene region. Oncol Rep 9:955–959.

Rebbeck TR, Sankar P [2005]. Ethnicity, ancestry, and race in molecular epidemiologic research. Cancer Epidemiol Biomarkers Prev 14:2467–2471.

Redon R, Ishikawa S, Fitch KR, Feuk L, Perry GH, Andrews TD, Fiegler H, Shapero MH, Carsons AR, Chen W, Cho, EK, Dallaire S, Freeman JL, Gonza´lez JR, Grataco`s M, Huang J, Kalaitzopoulos D, Komura D, MacDonald JR, Marshall CR, Mei R, Montgomery L, Nishimura K, Okamura K, Shen F, Somerville MJ, Tchinda J, Valsesia A, Woodwark C, Yang F, Zhang J, Zerjal T, Zhang J, Armengol L, Conrad DF, Estivill X, Tyler-Smith C, Carter NP, Aburatani H, Lee C, Jones KW, Scherer SW, Hurles ME [2006]. Global variation in copy number in the human genome. Nature *444*:444–454.

Renegar G, Reiser P, Manasco P [2001]. Family consent and the pursuit of better medicines through genetic research. J Contin Educ Health Prof 21:265–270.

Renegar G, Webster CJ, Stuerzebecher S, Harty L, Ide SE, Balkite B, Rogalski-Salter TA, Cohen N, Spear BB, Barnes DM, Brazell C [2006]. Returning genetic research results to individuals: points to consider. Bioethics *20*:24–36.

Richeldi L, Kreiss K, Mroz MM, Zhen B, Tartoni P, Saltini C [1997]. Interaction of genetic and exposure factors in the prevalence of berylliosis. Am J Ind Med *32*:337–340.

Richeldi L, Sorrentino R, Saltini C [1993]. *HLA-DPB1* glutamate 69: a genetic marker of beryllium disease. Science *262*:242–244.

Roberts LW, Geppert CMA, Warner TD, Hammond KAG, Rogers M, Smrcka J, Roberts BB [2005]. Perspectives on use and protection of genetic information in work settings: results of a preliminary study. Soc Sci Med *60*:1855–1858.

Robien K, Ulrich CM [2003]. *5,10-Methylenetetrahydrofolate reductase* polymorphisms and leukemia risk: a HuGE mini review. Am J Epidemiol *157*:571–582.

Rossman MD, Stubbs J, Lee CW, Argyris E, Magira E, Monos D [2002]. Human leukocyte antigen class II amino acid epitopes. Susceptibility and progression markers for beryllium hypersensitivity. Am J Respir Crit Care Med *165*:788–794.

Rossner P, Boffetta P, Ceppi M, Bonassi S, Smerhovsky Z, Landa K, Juzova D, Sram RJ [2005]. Chromosomal aberration in lymphocytes of healthy subjects and risk of cancer. Env Health Persp *113*:517–520.

Roth MJ, Dawsey SM, Wang G-Q, Tangrea JA, Zhou B, Ratnasinghe D, Woodson KG, Olivero OA, Poirier MC, Frye BL, Taylor PR, Weston A [2000]. Association between *GSTM1*0* and squamous dysplasia of the esophagus in the high-risk region of Linxian, China. Cancer Lett *156*:73–81.

Rothenberg K, Fuller B, Rothstein M, Duster T, Ellis Kahn MJ, Cunningham R, Fine B, Hudson K, King MC, Murphy P, Swergold G, Collins F [1997]. Genetic information and the workplace: legislative approaches and policy changes. Science *275*:1755–1757.

Rothman N [1995]. Genetic susceptibility biomarkers in studies of occupational and environmental cancer: methodologic issues. Toxicol Lett *77*:221–225.

Rothman KJ, Greenland S [1998]. Modern epidemiology. Philadelphia, PA: Lippincott-Raven.

Rothman N, Smith MT, Hayes RB, Traver RD, Hoener B, Campleman S, Li G-L, Dosemeci M, Linet M, Zhang L, Xi L, Wacholder S, Lu W, Meyer KB, Titentko-Holland N, Stewart JT, Yin S, Ross D [1997]. Benzene poisoning, a risk factor for hematological malignancy, is associated with the *NQO1 609*C>T mutation and rapid fractional excretion of chlorzoxazone. Cancer Res *57*:2839–2842.

Rothstein MA [2005]. Genetic exceptionalism and legislative pragmatism. Hastings Cent Rep *35*:27–33.

Rothstein MA, ed. [2003]. Pharmacogenomics: social, ethical, and clinical dimensions. Hoboken, NY: John Wiley & Sons, Inc.

Rothstein MA [2000a]. Ethical guidelines for medical research on workers. J Occup Environ Med 42(12):1166–1171.

Rothstein MA [2000b] Genetics and the Work Force of the Next Hundred Years. Colum Bus L Rev 3:371-402.

Rothstein MA [1984]. Genetic screening and monitoring. In: Medical screening of workers. Washington, DC: Bureau of National Affairs pp.52-61.

Rothstein MA, Knoppers BM [2005]. Introduction: regulation of biobanks. J Law Med Ethics 33:1–106.

Rystedt I [1985]. Hand eczema and long-term prognosis in atopic dermatitis. Acta Derm Venereol Suppl (Stockh) 117:1–59.

SACGT (Secretary's Advisory Committee on Genetic Testing) [2000]. Enhancing the Oversight of Genetic Tests: Recommendations of the SACGT. Bethesda, MD: National Institutes of Health. July. http://www4.od.nih.gov/oba/sacgt/gtdocuments.html. Date accessed: January 29, 2008.

Sadetzki S, Flint-Richter P, Starinsky S, Novikov I, Lerman Y, Goldman B, Friedman E [2005]. Genotyping of patients with sporadic and radiation-associated meningiomas. Cancer Epidemiol Biomarkers Prev 14:969–976.

Saltini C, Richeldi L, Losi M, Amicosante M, Voorter C, van den Berg-Loonen E, Dweik RA, Wiedemann HP, Duebner DC, Tinelli C [2001]. Major histocompatibility locus genetic markers of beryllium sensitization and disease. Eur Respir J 18:677–684.

Samuels SW [1998a]. Introduction: the Selikoff agenda. In: Samuels SW, Upton AC, eds. Genes, cancer, and ethics in the work environment. Beverly Farms, MA: OEM Press, pp. xvii–xxiv.

Samuels SW [1998b]. The Selikoff agenda and the human genome project: ethics and social issues. In: Samuels SW, Upton AC, eds. Genes, cancer, and ethics in the work environment. Beverly Farms, MA: OEM Press, pp. 3–9.

Sanderson S, Salanti G, Higgins J [2007]. Joint effects of the *N-acetyltransferase 1 and 2* (*NAT1* and *NAT2*) genes and smoking on bladder carcinogenesis: a literature-based systematic huge review and evidence synthesis. Am J Epidemiol 166:741-51.

Schill AL [2000]. Genetic information in the workplace: implications for occupational health surveillance. AAOHN J 48:80–91.

Schulte PA [2007]. The contribution of genetics and genomics to occupational safety and health. Occup Environ Med 64:717:718.

Schulte PA [2005]. The use of biomarkers in surveillance, medical screening, and intervention. Mutat Res 592:155–163.

Schulte PA [2004]. Some implications of genetic biomarkers in occupational epidemiology and practice. Scand J Work Environ Health 30:71–79.

Schulte PA [1987]. Simultaneous assessment of genetic and occupational risk factors. J Occup Med 29:884–891.

Schulte PA, DeBord DG [2000]. Public health assessment of genetic information in the occupational setting. In: Khoury MJ, Burke W, Thomson EJ, eds. Genetic and public health in the 21st century. New York: Oxford University Press, pp. 203–222.

Schulte PA, Halperin WE [1987]. Genetic screening and monitoring in the workplace. In: Harrington JM, ed. Recent advances in occupational health. Vol. 3. Edinburgh, Scotland, U.K.: Churchill Livingstone, pp. 135–154.

Schulte PA, Hunter D, Rothman N [1997]. Ethical and social issues in the use of biomarkers in epidemiological research. In: Toniolo P, Boffetta P, Shuker DEG, Rothman N, Hulka B, Pearce N, eds. Application of biomarkers in cancer epidemiology IARC Scientific Publication No. 142. Lyon, France: International Agency for Research on Cancer, pp. 313–318.

Schulte PA, Lomax G [2003]. Assessment of the scientific basis for genetic testing of railroad workers with carpal tunnel syndrome. J Occup Environ Med 45:592–600.

Schulte PA, Lomax GP, Ward EM, Colligan MJ [1999]. Ethical issues in the use of genetic markers in occupational epidemiologic research. J Occup Environ Med 41:639–646.

Schulte PA, Perera FP [1997]. Transitional studies. In: Toniolo P, Boffetta P, Shuker DEG, Rothman N, Hulka B, Pearce N, eds. Application of biomarkers in cancer epidemiology. IARC Scientific Publication No. 142. Lyon, France: International Agency for Research on Cancer, pp. 19–29.

Schulte PA, Perera FP [1993]. Validation. In: Schulte PA, Perera FP, eds. Molecular epidemiology: principles and practices. San Diego, CA: Academic Press pp.79–107.

Schulte PA, Talaska G [1995]. Validity criteria for the use of biological markers of exposure to chemical agents in environmental epidemiology. Toxicology 101:73–88.

Segal E, Friedman N, Kaminski J, Regev A, Koller D [2005]. From signatures to models: understanding cancer using microarrays. Nat Genet 37:S-38-S45.

Seminara D, Khoury MJ, O'Brien TR, Manolio T, Gwinn ML, Little J, Higgins JP, Bernstein JL, Boffetta P, Bondy M, Bray MS, Brenchley PE, Buffler PA, Casas JP, Chokkalingam AP, Danesh J, Davey Smith G, Dolan S, Duncan R, Gruis NA, Hashibe M, Hunter D, Jarvelin MR, Malmer B, Maraganore DM, Newton-Bishop JA, Riboli E, Salanti G, Taioli E, Timpson N, Uitterlinden AG, Vineis P, Wareham N, Winn DM, Zimmern R, Ioannidis JP; Human Genome Epidemiology Network; the Network of Investigator Networks [2007].

The emergence of networks in human genome epidemiology: challenges and opportunities. Epidemiology *18*:1-8.

Sexton K, Balharry D, BeruBe KA [2008]. Genomic biomarkers of pulmonary exposure to tobacco smoke components. Pharmacogenet Genomics *18*:853–860.

Sharp L, Cardy AH, Cotton SC, Little J [2004]. *CYP17* gene polymorphisms: prevalence and associations with hormone levels and related factors: a HuGE review. Am J Epidemiol *160*:729–740.

Sharp L, Little J [2004]. Polymorphisms in genes involved in folate metabolism and colorectal neoplasia: a HuGE review. Am J Epidemiol *159*:423–443.

Shostak S [2003]. Locating gene-environment interaction: at the intersections of genetics and public health. Soc Sci Med *56*:2327–2342.

Sobti RC, Sharma S, Joshi A, Jindal SK, Janmeja A [2004]. Genetic polymorphism of the *CYP1A1, CYP2E1, GSTM1* and *GSTT1* genes and lung cancer susceptibility in a north Indian population. Mol Cell Biochem *266*:1–9.

Spiridonova MG, Stepanov VA, Puzyrev VP, Karpov RS [2001]. The estimation of gametic disequilibrium between DNA markers in candidate genes for coronary artery disease (CAD) and the associations of gene complexes with risk factors for CAD. Int J Circumpolar Health *160*:222–227.

Staley K [2003]. Genetic testing in the workplace: a report by GeneWatch UK. http://www.genewatch.org/uploads/f03c6d66a9b354535738483c1c3d49e4/GeneticTesting.pdf. Date accessed: June 4, 2007.

Stokinger HE, Mountain JT [1963]. Test for hypersusceptibility to hemolytic chemicals. Arch Environ Health *6*:495–502.

Strudler A [1994]. The social construction of genetic abnormality: ethical implications for managerial decisions in the workplace. J Bus Ethics *13*:839–848.

Stumvoll M, Goldstein BJ, van Haeften TW [2005]. Type 2 diabetes: principles of pathogenesis and therapy. Lancet *365*:1333–1346.

Taioli E, Benhamou S, Bouchardy C, Cascorbi I, Cajas-Salazar N, Dally H, Fong KM, Larsen JE, Le Marchand L, London SJ, Risch A, Spitz MR, Stucker I, Weinshenker B, Wu X, Yang P [2007]. *Myeloperoxidase G-463A* polymorphism and lung cancer: a HuGE genetic susceptibility to environmental carcinogens pooled analysis. Genet Med *9*:67–73.

Taioli E, Sram RJ, Binkova B, Kalina I, Popov TA, Garte S, Farmer PB [2007]. Biomarkers of exposure to carcinogenic PAHs and their relationship with environmental factors. Mutat Res *620*:16-21.

Takaro TK, Engel LS, Keifer M, Bigbee WL, Kavanagh TJ, Checkoway H [2004].

Glycophorin A is a potential biomarker for the mutagenic effects of pesticides. Int J Occup Environ Health *10*:256–261.

Tamburro CH, Wong JL [1993]. Practical applications of biomarkers in the study of environmental liver disease. In: Schulte PA, Perera FP, eds. Molecular epidemiology: principles and practices. San Diego, CA: Academic Press, pp. 517–546.

Taubes G [1995]. Epidemiology faces its limits. Science *269*:164–169.

Taylor AN [2001]. Role of human leukocyte antigen phenotype and exposure in development of occupational asthma. Curr Opin Allergy Clin Immunol *1*:157–161.

Teixeira JP, Gaspar J, Silva S, Torres J, Silva SN, Azevedo MC, Neves P, Laffon B, Mendez J, Goncalves C, Mayan O, Farmer PB, Rueff J [2004]. Occupational exposure to styrene: modulation of cytogenetic damage and levels of urinary metabolites of styrene by polymorphisms in genes *CYP2E1*, *EPHX1*, *GSTM1*, *GSTT1* and *GSTP1*. Toxicology *195*:231–242.

Teixeira JP, Silva S, Torres J, Gaspar J, Roach J, Farmer PB, Rueff J, Mayan O [2008]. Styrene-oxide N-terminal valine haemoglobin adducts as biomarkers of occupational exposure to styrene. Int J Hyg Environ Health *211*:59–62.

Tennant RW [2002]. The National Center for Toxicogenomics: using new technologies to inform mechanistic toxicology. Environ Health Perspect *110*:A8–A10.

Thier R, Bruning T, Roos PH, Rihs H-P, Golka K, Bolt HM [2003]. Markers of genetic susceptibility in human environmental hygiene and toxicology: the role of selected SYP, NAT, and GST genes. Int J Hyg Environ Health *206*:149–171.

Tilton SH [1996]. Right to privacy and confidentiality of medical records. Occup Med *11*:17–29.

Tinkle SS, Weston A, Flint MS [2003]. Genetic factors modify the risk of developing beryllium disease. Semin Respir Crit Care Med 24:169–178.

Toraason MA, Hayden C, Marlow D, Rinehart R, Mathias P, Werren D, DeBord DG, Reid TM [2001]. DNA strand breaks, oxidative damage, and 1-OH pyrene I roofers with coal tar pitch dust and /or asphalt fume exposure. Int Arch Occup Environ Health *74*:396–404.

Toyoshiba H, Yamanaka T, Sone H, Parham FM, Walker NJ, Martinez J, Portier CJ [2004]. Gene interaction network suggests dioxin induces a significant linkage between aryl hydrocarbon receptor and retinoic acid receptor beta. Environ Health Perspect *112*:1217–1224.

Travis CC, Bishop WE, Clark DP [2003]. The genomic revolution: what does it mean for human and ecological risk assessment? Ecotoxicology *12*:489-495.

Truett J, Cornfield J, Kannel W [1967]. A multivariate analysis of the risk of coronary heart disease in Framingham. J Chronic Dis *20*:511–524.

Uniform Health Care Information Act [1986]. http://aspe.hhs.gov/admnsimp/PVCREC3.HTM. Date accessed: August 29, 2007.

U.S. Congress [2008]. Genetic Information Nondiscrimination Act. Text of legislation S358: http://thomas.loc.gov/cgi-bin/bdquery/z?d110:S.358: ; Text of house bill 493: http://thomas.loc.gov/cgi-bin/bdquery/z?d110:h.r.00493:; Date accessed: May 21, 2008.

U.S. Task Force on Genetic Testing [1998]. Promoting safe and effective genetic testing in the United States: final report pf the task force on genetic testing. Holtzman NA, Watson MS, eds. Baltimore, MD: John Hopkins Press.

Van Damme K, Casteleyn L [2003]. Current scientific, ethical and social issues of biomonitoring in the European Union. Toxicol Lett *144*:117–126.

Van Damme K, Casteleyn L, Heseltine E, Huici A, Sorsa M, van Larebeke N, Vineis P [1995]. Individual susceptibility and prevention of occupational diseases: scientific and ethical issues. J Occup Environ Med *37*:91–99.

Vineis P [2007]. Methodological approaches to gene-environment interactions in occupational epidemiology. Occup Environ Med *64*:e3.

Vineis P [1992]. Uses of biochemical and biological markers in occupational epidemiology. Rev Epidemiol Sante Publique *40*:S63–S69.

Vineis P, Caporaso N, Tannenbaum, SR, Skipper PL, Glogowski J, Bartsch H, Coda M, Talaska G, Kadlubar F [1990]. Acetylation phenotype, carcinogen-hemoglobin adducts, and cigarette smoking. Cancer Res *50*:3002–3004.

Vineis P, Malats N, Lang M, d'Errico A, Caporaso N, Cuzick J, Boffetta P, eds. [1999]. Metabolic polymorphisms and susceptibility to cancer. IARC Scientific Publication No. 148. Lyon, France: International Agency for Research on Cancer.

Vineis P, Schulte PA, McMichael AJ [2001]. Misconceptions about the use of genetic tests in populations. Lancet *357*:709–712.

Vineis P, Schulte PA, Vogt RF Jr. [1993]. Technical variability in laboratory data. In: Schulte PA, Perera FP, eds. Molecular epidemiology: principles and practices. San Diego, CA: Academic Press, pp. 109–135.

Wade PA, Archer TK [2006]. Epigenetics: environmental instructions for the genome. Environ Health Perspect *114*:A140–A141.

Wald N, Cuckle H [1989]. Reporting the assessment of screening and diagnostic tests. Br J Obstet Gynaecol *96*:389–396.

Walston J, Silver K, Bogardus C, Knowler WC, Cell FS, Austin S, Manning B, Strosberg AD, Stern MP, Raben N, Sorkin JD, Roth J, Shuldiner AR [1995]. Time of onset of non-insulin-dependent diabetes mellitus and genetic variation in the ß3-adrenergic receptor gene. N Engl J Med *333*:343–347.

Wang G-Y, Lu C-Q, Zhang RM, Hu XH, Luo ZW [2008]. The *E-cadherin* Gene Polymorphism -160C→A and Cancer Risk: A HuGE Review and Meta-Analysis of 26 Case-Control Studies. Am J of Epidem *167*:7-14.

Wang WY, Barrett BJ, Clayton DG, Todd JA [2005]. Genome-wide association studies: theoretical and practical concerns. Nat Rev Genet *6*:109-118.

Wang Z, Farris GM, Newman LS, Shou Y, Maier LA, Smith HN, Marrone BL [2001]. Beryllium sensitivity is linked to HLA-DP genotype. Toxicology *165*:27–38.

Wang Z, Neuburg D, Li C, Su L, Kim JY, Chen JC, Christiani DC [2005]. Global gene expression profiling in whole-blood samples from individuals exposed to metal fumes. Env Health Perspect *113*:233–241.

Wang Z, White PS, Petrovic M, Tatum OL, Newman LS, Maier LA, Marrone BL [1999]. Differential susceptibilities to chronic beryllium disease contributed by different Glu^{69} HLA-DPB1 and -DPA1 alleles. J Immunol *163*:1647–1653.

Ward EM, Hurrell JJ, Colligan MJ [2002]. Ethical issues in occupational health research. Occup Med *17*:637–655.

Waters MD, Olden K, Tennant RW [2003]. Toxicogenomic approach for assessing toxicant-related disease. Mutat Res *544*:415–424.

Watson MS, Greene CL [2001]. Points to consider in preventing unfair discrimination in genetics disease risk: a position statement of the American College of Medical Genetics. Genet Med *3*:436–437.

Weeks JL, Levy BS, Wagner GR [1991]. Preventing occupational disease and injury. Washington DC: American Public Health Association, pp. 45–51.

Weinhold B [2006]. Epigenetics:the science of change. Environ Health Perspect *114*:A160-A167.

Weir BA, Woo MS, Getz G, Perner S, Ding L, Beroukhim R, Lin WM, Province MA, Kraja A, Johnson LA, Shah K, Sato M, Thomas RK, Barletta JA, Borecki IB, Broderick S, Chang AC, Chiang DY, Chirieac LR, Cho J, Fujii Y, Gazdar AF, Giordano T, Greulich H, Hanna M, Johnson BE, Kris MG, Lash A, Lin L, Lindeman N, Mardis ER, McPherson JD, Minna JD, Morgan MB, Nadel M, Orringer MB, Osborne JR, Ozenberger B, Ramos AH, Robinson J, Roth JA, Rusch V, Sasaki H, Shepherd F, Sougnez C, Spitz MR, Tsao MS, Twomey D, Verhaak RG, Weinstock GM, Wheeler DA, Winckler W, Yoshizawa A, Yu S, Zakowski MF, Zhang Q, Beer DG, Wistuba II, Watson MA, Garraway LA, Ladanyi M, Travis WD, Pao W, Rubin MA, Gabriel SB, Gibbs RA, Varmus HE, Wilson RK, Lander ES, Meyerson M [2007]. Characterizing the cancer genome in lung adenocarcinoma. Nature *450*:893-898.

Weiss JR, Moysich KB, Swede H [2005]. Epidemiology of male breast cancer. Cancer Epidemiol Biomarkers Prev *14*:20–26.

Wen W, Che W, Lu L, Yang J, Gao X, Wen J, Heng Z, Cao S, Cheng H [2008]. Increased damage of exon 5 of p53 gene in workers from an arsenic plant. Mutat Res *643*:36–40.

Wenzlaff AS, Cote ML, Bock CH, Land SJ, Schwartz AG [2005]. *GSTM1, GSTT1*, and *GSTP1* polymorphisms, environmental tobacco smoke exposure and risk of lung cancer among never smokers: a population-based study. Carcinogenesis *26*:395–401.

Weston A, Ensey J, Kreiss K, Keshava C, McCanlies EC [2002]. Racial differences in prevalence of a supratypic HLA-genetic marker immaterial to pre-employment testing for susceptibility to chronic beryllium disease. Am J Ind Med *41*:457–465.

White DL, Li D, Nurgalieva Z, El-Serag HB [2008]. Genetic variants of *glutathione s-transferase* and possible risk factors for hepatocellular carcinoma: a huge systematic review and meta-analysis. Am J Epidemiol. *167*:377–389.

Wiesner G [1997]. Clinical implications of BRCA 1 genetic testing in Ashkenazi-Jewish women. Health Matrix Clevel *7*:3–30.

Wikman H, Piirila P, Rosenberg C, Luukkonen R, Kaaria K, Nordman H, Norppa H, Vainio H, Hirvonen A [2002]. N-Acetyltransferase genotypes as modifiers of diisocyanate exposure-associated asthma risk. Pharmacogenetics *12*:227–233.

Williams MS [2001]. Genetics and managed care: policy statement of the American College of Medical Genetics. Genet Med *3*:430–435.

Wittes J, Friedman HP [1999]. Searching for evidence of altered gene expression: a comment on statistical analysis of microarray data. J Natl Cancer Inst *91*:400–401.

Workers' Compensation Statutes [2007]. http://www.law.cornell.edu/topics/workerscompensation.html. Date accessed: February 20, 2008.

Wu F, Tsai F, Kuo H, Tsai C, Wu W, Wang R, Lai J [2000]. Cytogenetic study of workers exposed to chromium compounds. Mutat Res *464*:289–296.

Xia Y, Bian Q, Xu L, Cheng S, Song L, Liu J, Wu W, Wang S, Wang X [2004]. Genotoxic effects on human spermatozoa among pesticide factory workers exposed to fenvalerate. Toxicology *203*:49–60.

Yang Q, Khoury MJ, Coughlin SS, Sun F, Flanders WD [2000]. On the use of population-based registries in the clinical validation of genetic tests for disease susceptibility. Genet Med *2*:186–192.

Yesley MS [1999]. Genetic difference in the workplace. Jurimetrics *40*:129–142.

Yong LC, Schulte PA, Wiencke J, Boeniger MF, Connally LB, Walker JT, Whelan EA, Ward EM [2001]. Hemoglobin adducts and sister chromatid exchanges in hospital workers exposed to ethylene oxide: effects of *glutathione S-transferase T1* and *M1* genotypes. Cancer Epidemiol Biomarkers Prev *10*:539–550.

Yucesoy B, Luster MI [2007]. Genetic susceptibility in pneumoconiosis. Toxicol Lett *168*:249–254.

Yucesoy B, Vallyathan V, Landsittel DP, Sharp DS, Weston A, Burleson GR, Simeonova P, McKinstry M, Luster MI [2001]. Association of *tumor necrosis factor*-α and *interleukin-1* gene polymorphisms with silicosis. Toxicol Appl Pharmacol *172*:75–82.

Web sites

WEB SITES FOR FURTHER INFORMATION

American College of Medical Genetics. http://www.acmg.net/

American College of Occupational and Environmental Medicine (ACOEM). http://www.acoem.org/

American Society of Human Genetics (ASHG). http://www.ashg.org

Environmental Genome Project (EGP). http://www.niehs.nih.gov/research/supported/programs/egp/index/htm

European Group on Ethics of Science and New Technology: Ethical Aspects of Genetics Testing in the Workplace. http://ec.europa.eu/european_group_ethics/index_en.htm

Fred Hutchinson Cancer Research Center Quantitative Genetic Epidemiology http://cougar.fhcrc.org/

Genetic Epidemiology Group. http://www.niehs.nih.gov/research/atniehs/labs/lrb/gen-epi/index.cfm

Genetics and Public Policy Center. http://www.dnapolicy.org/

Genome Programs of the U.S. Department of Energy Office of Science. http://www.doegenomes.org/

HuGENet. http://www.cdc.gov/genomics/hugenet/about.htm

Laboratory of Molecular Genetics http://www.niehs.nih.gov/research/atniehs/labs/lmg/index.cfm

National Center for Toxicogenomics. http://www.niehs.nih.gov/nct/home.htm

National Human Genome Research Institute (NHGRI). http://www.nhgri.nih.gov

National Institute for Occupational Safety and Health. http://www.cdc.gov/niosh/homepage.html

Occupational Safety and Health Administration (OSHA). http://www.osha.gov

National Office of Public Health Genomics. CDC. http://www.cdc.gov/genomics/

Office of Technology Assessment (OTA). Genetic monitoring and screening in the workplace. Washington, DC: Congress of the United States, OTA–B4–455, 1990. http://www.princeton.edu/~ota/ns20/alpha_f.html

Workers' Compensation Statutes. Date accessed: February 2, 2004. http://www.law.cornell.edu/topics/workers_compensation.html

Index

ACCE.. 19
Acquired genetic effects 5, 12–13, 15, 62, 63
    Definition of 5
        Practice 14
        Regulation and litigation 75–76
        Research 74–75
Adduct.. . . . . 2, 15, 23,74
    DNA, definition of xvi
    Protein, definition of xx
Allele... . . . . 6, 22, 29, 32, 36, 37
    Definition of.. xv
Americans with Disability Act of 1990 (ADA) 40–42, 57
Asthma, occupational induced.. 29
Autonomy 41, 63, 69, 70–71
Banked or stored specimens. 35, 65, 75
Base pair, definition of. xv
Bioinformatics.. . . . .24, 75
    Definition of. xv
Biologically effective dose 1–2, 3, 14
    Definition of. xv
Biomarker 1–4
    Definition of. xv
    Effective of genes and use of risk assessments. 14
    Framework for consideration 1–4
    Genetic damage. 13, 15, 47–52
    Types of. 1–4
        Disease 2
        Exposure 1–2, 3
        Susceptibility. 2
    Validation of. 22–23
Cell lines 35–36
Centromere, definition of. xv
Chemical base, definition of xv
Chromosomal aberration, definition of. xv
Chromosome, definition of.. xv
Chronic beryllium disease. 4, 22–23, 56
Confidentiality.. . . . . 40–42, 70–71
    Regulation.. 41–42
Confounding 30, 34, 36, 37, 47, 49, 62
    Definition of. xvi
Copy number variant (CNV) 7

Definition of..xvi
Deoxyribonucleic acid (DNA)
    Adduct, definition of...xvi
    Definition of..xvi
    Strand break, definition of...xvi
Diagnostic test..19, 20, 25
    Definition of..xvi
Discrimination..40, 43, 68
Effect of exposures on genetic material...12–13
Effect modification...27
    Definition of..xvi
Electrophile, definition of...xvi
Epigenetics...13
    Definition of..xvi
Environmental Genome Project..17
Ethical, Legal and Social Issues
    Autonomy...41, 63, 69, 70–71
    Discrimination..40, 68
    Economics..61–62
    Genetic information in research..62-66, 74–75
    Genetics in workplace...61–79
    Susceptibility..62–72
Evaluation of genomic applications in practice and prevention (EGAPP)..............66
Gene
    Definition of..xvi
Gene-environment interaction..5–12, 27–30
    Definition of..xvi
    In occupational safety and health research...5–12
    Issues..5-7, 27-30, 32–34
    New technologies..24–27
    Relationship with diseases..28
        Silicosis...12
        Smoking..14
    Study design and analysis..32–34, 36–37
Gene-gene interactions, definition of..xvii
Genetic assays/tests
    Clinical Laboratory Improvement Amendment of 1988 (CLIA).................67
    Validity and utility issues..19–23, 67
Genetic discrimination, definition of...xvii
Genetic exceptionalism..42–43
    Definition of..xvii

Genetic information
- Adequacy of safeguards to prevent misuse... 78–79
- Assessment of causation ... 45, 72–73
- Communication and interpretation of results... 17–18, 77
- Considerations... 32–34
- Definition of... xvii
- Framework consideration... 62
- Human Epidemiology Genetic Epidemiology Network (HuGENet)... 30–31
- Historical use ... 43–44
- Incorporation in health records ... 39–45
- Misuse of... 70–73
- Policy statements ... 41–42
- Regulation... 40–41, 72–73, 75–77
- Role in occupational disease... 5–15, 27–35
- Source of in health records ... 39–45
- Use in research ... 44
- Use in the workplace... 39–40, 66–72, 75

Genetic Information Nondiscrimination Act of 2008 (GINA) ... 40, 50, 68

Genetic monitoring
- Considerations... 51–52
- Definition of ... xvii
- Historical use... 47–50
  - Chernobyl cleanup workers ... 48
  - Surveillance programs ... 47–49
- In research... 74–75
- In practice ... 47–52, 75
  - In risk assessment ... 49–50
- Regulation... 50–51

Genetic research,
- Analytical epidemiology ... 30–35
- Anonymization ... 65
- Communicating results... 65–66
- Considerations for the incorporation into occupational health... 17–37
  - Checklist... 32–34
- Definition of... xvii
- Interpreting results... 65–66
- Prevention and diagnosis... 66–67
- Regulations and guidance policies ... 63, 72–73, 75–77
- Safeguarding rights of participants... 62–65
- Study design... 36–37
- Types of... 3
- Validation of susceptibility markers ... 22–23

Genetic screening
    Application of 57–59
    Communication and interpretation 65–66
    Criteria for 57–59
    Definition of xvii
    Economic issues 61–62
    Ethical implications 67–70
    Hierarchy of prevention 57
    History 54–55
    Informed consent 64–65
    In job actions 67–70
    In research 62–66
    Objectives of 55–57
    Past and current use 55–57
    Public health issues 57–58, 67–70
    Regulations and guidance policies 57–58, 71–73
    Technical issues 57
    Theoretical use in practice 53–59
Genetic susceptibility, inclusion in risk assessment 14, 73–74
Genetic test
    Definition of xvii
    Informed consent 64–65
    Use in disease diagnosis 66–67
Genetic testing
    Current and past use 43–45, 55–57
        Chernobyl cleanup workers 48
    Litigation 72–73
    Regulations and guidance policies 74
Genome, definition of xvii
Genome-wide association studies, definition of xviii
Genomics,
    Definition of xviii
    Priorities in occupational safety and health 14
    Challenges and related research areas 24–27
Genotype, definition of xviii
Hardy-Weinberg equilibrium, definition of xviii
Health records, Source of genetic information 39–45
Health inquiries and examinations 40
Heterozygous, definition of xviii
Hierarchy of controls 49–50, 57
    Definition of xviii
Homozygous definition of xviii

Human Genetic Epidemiology Network (HuGENet) 30–31
    Reviews. 31
Inherited genetic factors
    Definition of. 5
        Regulation and litigation 72–74
        Research 62–66
        Practice.. 66–72
Institutional Review Board (IRB).. .. ..44, 79
Linkage disequilibrium, definition of xviii
Litigation
    Acquired genetic factors 75–77
    Inherited genetic factors. 72–74
Locus, definition of xviii
Medical monitoring. 47–50
    Definition of. xviii
Medical removal 47–50
    Definition of. xviii
Medical screening, definition of xviii
Medical surveillance, definition of xix
Metabonomics.. 24,74
    Definition of. xix
Microarray
    Challenges.. 24–27
    Definition of. xix
    DNA. 24–27
    Issues 25–27
    Protein 24–27
    Statistical approaches 24–27
    Utility of 24–27
Micronuclei, definition of xix
Monogenic disease, definition of xix
Mutation, definition of xix
Negative Predictive Value, definition of.. xix
Network of Networks, ... 35
Occupational health research, incorporating genetic information 17–35
Occupational health practice
    Genetic monitoring 47–52, 75
    Role of polymorphisms.. .. .. 53–59, 66–72
Oncogene, definition of xix
Outlier, definition of xix
Penetrance 21
    Definition of.. xix

Phase I enzymes, definition of ... xix
Phase II enzymes, definition of ... xix
Phenotype, definition of ... xix
Polygenic disease, definition of ... xix
Polymerase chain reaction, definition of ... xx
Polymorphism
    Definition of ... xx
    HapMap ... 7
    Relationship to variability ... 5–12
Positive predictive value (PPV), definition of ... xx
Privacy ... 40–42, 70–71
Protein expression, definition of ... xx
Proteome, definition of ... xx
Proteomics, definition of ... xx
Protooncogene, definition of ... xx
Regulation and policy statements
    Acquired genetic factors ... 75–77
    Confidentiality ... 40–42
    Genetic discrimination ... 43, 78–79
    Genetic information ... 41–42, 72–74, 75–78
    Genetic monitoring ... 50–51
    Genetic testing ... 57–58
    Inherited genetic factors ... 73
Reporter gene ... 15
    Definition of ... xx
Research, analytical epidemiological ... 30–35
Resequencing, definition of ... xx
Restriction enzymes, definition of ... xx
Restriction fragment length, definition of ... xxi
Ribonucleic acid (RNA), definition of ... xxi
Risk assessment ... 75
Risk factors, relationship between genetic and environmental ... 27–30
Security ... 40–42
Sensitivity, definition of ... xxi
Sequencing, definition of ... xxi
Single nucleotide polymorphism (SNP), definition of ... xxi
Sister chromatid exchange, definition of ... xxi
Sister chromatids, definition of ... xxi
Social justice ... 40
Specificity, definition of ... xxi

Stigmatization..71–72
Stochastic process, definition of..xxi
Susceptibility..5–7, 53–59, 61–62
    Definition of...xxi
Technical variability, definition of..xxi
Toxicogenomics..24, 75
    Definition of...xxii
Transcriptomics, definition of..xxii
Transcriptosome, definition of...xxii
Transgenic animal...35–36
    Definition of...xxii
Transitional study..1–4
    Definition of...xxii
Utility,
    Clinical,...21–22
        Definition of...xxii
    Issues of genetic assays..19–23
Validation,
    Definition of...xxii
    Evaluation of exposure and genetic damage................................22
    Evaluation of genetic polymorphisms.....................................22–23
    Evaluation of multiple biomarkers...23
    Sources of error and bias...23
Validity
    Analytical..19–21
        Definition of...xv
    Clinical, definition of,..xvi
    Issues of genetic assays..19–23
Variability
    Biological...xv, 5–6
    HapMap Project..7
    Interindividual..5–12
    Intraindividual...5–12
        Response..5–12
        Sources of..5–12
    Xenobiotic, definition of..xxii

www.ingramcontent.com/pod-product-compliance
Lightning Source LLC
Chambersburg PA
CBHW080254180526
45167CB00006B/2531